Mind-Blowing Connections For Kids! The Book of Interesting, Weird -But-True Fun Science Facts

Why Earth, Space, Plants, Animals & The Brains of Smart, Curious Children Share Surprising Links

Keryn Gold, PhD, MBA

Lead With Gold

Author: Keryn Gold, PhD, MBA
Publisher: Lead With Gold, LLC

Legal Disclaimer

This publication is intended to provide accurate and authoritative information regarding the subject matter covered. However, it is sold with the understanding that neither the author nor the publisher is engaged in rendering legal, financial, investment, medical, accounting, or other professional services.

While every effort has been made to ensure the accuracy and completeness of the information in this book, the author and publisher make no guarantees, express or implied, regarding its applicability, reliability, or effectiveness for your individual circumstances. The strategies and advice herein may not be suitable for all individuals or organizations. Readers should seek professional guidance where appropriate.

Neither the publisher nor the author shall be held liable for any direct, indirect, incidental, consequential, or other damages arising from the use or reliance on this book's content, including but not limited to financial loss, legal disputes, or personal and professional consequences.

Responsibility & Risk

By reading this book, you acknowledge that:

- You are solely responsible for your decisions, actions, and outcomes.
- The author and publisher do not accept liability for any financial, legal, or business results based on the material provided.
- Any case studies, examples, and statistics are either drawn from publicly available data or represent composite stories from the author's clients and experience to protect confidentiality.

Legal Notice

Paperback ISBN: 978-1-967456-13-0
Hardcover ISBN: 978-1-967456-14-7
eBook ISBN: 978-1-967456-12-3
Printed in the United States of America
First Edition, 2025
LeadWithGold.com

Contents

INTRODUCTION

— Everything is Connected!

Welcome to **Mind-Blowing Connections For Kids!** You are about to explore a world full of **surprising links** between **Earth, space, plants, animals**, and even **your own brain**. Have you ever wondered how **space dust** can end up in your backyard, or why **volcanoes** can change the color of the sky on the other side of the planet? In these pages, you'll learn amazing truths about how everything around us is linked in ways you never imagined.

As you read this book, get ready for **weird-but-true science facts** that will spark your curiosity. You'll discover how **parrot-fish poop** makes sand for tropical beaches, why **ants protect tiny bugs** to get a sweet reward, and how **trees talk to each other** using fungi underground. Every chapter will show you how living things—and even nonliving things—are part of **one big connected family** on our planet (and beyond!).

Together, we'll dive into **50 short and exciting chapters** filled with fun, easy-to-understand science. Each chapter includes a fascinating story, cool facts, and kid-friendly explanations for anyone who loves exploring and learning! So get comfy, flip the page, and let the adventure begin. We hope this book will tickle your brain and inspire you to keep asking **"Why?"** long after you've read the last page. Enjoy the journey!

1

The Great Space Dust Mystery

—How Tiny Specks from the Sky End Up in Your Backyard

Have you ever looked up at the night sky and wondered what's really out there? Stars, planets, and comets twinkle overhead, but did you know that tiny specks of **space dust** also fall from the sky every single day? Even more amazing, some of those **cosmic grains** may end up in your very own backyard!

Where Does Space Dust Come From?

Space dust comes from many places in our solar system. Sometimes it's leftover bits of rock and ice from **comets** that sweep past Earth. Other times, it's little pieces that break off of **asteroids** or **meteorites**. As these bits zoom through space and enter

Earth's atmosphere, they burn up or slow down, eventually drifting to the ground. By the time they land, many are almost invisible—like tiny, shiny specks you might not even notice.

How Can Dust Travel So Far?

Think about how a gentle breeze can blow a feather around. Space dust may be small, but it's super tough! It can zoom through space at **thousands of miles per hour** before slowing down and floating to Earth. If you have a microscope and a keen eye, you might spot strange little bits of metal or rock mixed in with regular dust on your windowsill or in a rain gutter. Some of those could be **dust from outer space**!

Why Is This Dust Important?

- **Clue to Earth's Past:** Scientists study space dust to learn about the early days of our solar system—before Earth was even fully formed!

- **Connection to Comets and Asteroids:** These little specks are like a **postcard** from faraway asteroids and comets, telling us what those distant objects are made of.

- **Help for Plants:** Microscopic dust from **meteorites** can sometimes contain minerals that sprinkle into the soil. Over time, these minerals can help plants grow.

Did You Know?

- Every year, **over 40,000 tons** of cosmic dust drift down onto Earth! That's heavier than 200 jumbo jets put together.

- Some people collect **micrometeorites** (tiny meteor fragments) from their rooftops. They use magnets or special filters to separate them from regular dirt.

Try It Yourself

1. **Collect Dust:** Place a white sheet of paper in an outdoor area that's protected from wind and rain (like a balcony or porch). Leave it out for a few days.

2. **Use a Magnet:** Gently wave a magnet around over the dust collected on your paper. If tiny specks stick to the magnet, they might be bits of **meteorite** or other metal fragments!

3. **Look Closely:** If you have a magnifying glass or microscope, see if any grains look different from normal dirt—maybe they're shiny or spherical. You never know what you might find.

Why It Matters

Space dust is a perfect example of how our lives on Earth connect to the farthest parts of our solar system. These tiny particles travel incredible distances, crossing **millions of miles** of emptiness to end up on our planet—and maybe in your very own neighborhood. It's one more reminder that we're all part of a **huge cosmic ecosystem**, where even the smallest speck has a story to tell!

2

Fish That Make Sand

– *The Funny Truth About Parrotfish Poop & The Sand You Walk On At the Beach*

Have you ever walked along a beach and felt the soft sand between your toes? Believe it or not, **some of that sand** may have come from a fish's tummy—specifically, a bright, colorful fish called the **parrotfish**. These unique reef dwellers have a hilarious (and helpful) secret: they use their powerful "beaks" to nibble on coral, then grind it up in their bellies. When they're done digesting, they poop out what becomes **fresh sand** for tropical beaches!

How Do Parrotfish Make Sand?

1. **Chomp, Chomp!** Parrotfish have strong, parrot-like

mouths that can scrape off tiny pieces of coral.

2. **Munch and Crush:** Inside their bellies, special teeth grind the coral into teeny, tiny particles.

3. **Sandy Surprise:** The leftover bits are then pooped out, turning into sand on the ocean floor. Ocean currents eventually wash it onto shore, where we enjoy it as powdery beach sand.

Why Is This Important?

- **Beach Builders:** Parrotfish poop plays a huge role in creating and maintaining those beautiful white beaches that tourists and sea animals both love.

- **Healthy Reefs:** By nibbling on algae and coral, parrotfish prevent reefs from getting clogged with too much algae. This helps coral reefs stay healthy and full of life.

- **Ecosystem Hero:** Without parrotfish, many coral reefs could suffer, and the soft white sand we enjoy would be a lot less plentiful.

Fun Facts

- A single parrotfish can produce **hundreds of pounds** of sand in just one year!
- Parrotfish have bright colors—blue, green, pink, or yellow—making them look like swimming rainbows.
- Some parrotfish even form a "mucus bubble" around themselves at night, like a sleeping bag, to hide their scent from predators.

Try It Yourself

1. **Ocean Detective:** Next time you visit a beach, scoop up some sand and look closely. Some tiny grains might have once been **coral** inside a parrotfish's tummy!
2. **Draw a Parrotfish:** Create your own colorful parrotfish art. Give it a bright beak and sparkly scales.

Why It All Matters

Parrotfish may seem silly with their goofy beaks and sandy poop, but they're superstars for our beaches and coral reefs. By making sand and keeping reefs clean, they show us how even **odd animal habits** can be a vital link in our planet's grand puzzle!

3

Clouds That Travel the World

– *Raindrops Here Today, Gone Tomorrow On A Globe-Trotting Adventure!*

Have you ever watched big, fluffy **clouds** drifting across the sky? It might look like they're just floating around without a plan, but these clouds are part of an **incredible journey**! Every cloud you see is carrying **tiny drops of water** that travel from oceans to forests, from deserts to mountains—and sometimes back again. It's like a **water highway** in the sky, linking every corner of the Earth!

One Giant Water Cycle

- **Evaporation:** The Sun warms lakes, oceans, and puddles, causing water to **evaporate** (turn into invisible water vapor) and rise into the air.

- **Condensation:** High in the sky, water vapor cools and **condenses** into tiny droplets. Those droplets stick together to form clouds.
- **Precipitation:** When the droplets get heavy, **rain** or **snow** falls from the clouds back to Earth, where the cycle starts all over again.

Why Is This Important?

- **Connecting Continents:** Clouds can carry water droplets for **hundreds or even thousands of miles**, meaning a raindrop that falls in your hometown might have formed over a distant ocean!
- **Helping Plants Grow:** Rain from clouds provides much-needed water for trees, flowers, and crops. Without traveling clouds, many places would be too dry to support life.
- **Cooling the Earth:** Clouds help **regulate tempera-**

ture by reflecting some of the Sun's heat away from Earth, like a giant umbrella in the sky.

Fun Facts

- A typical **cumulus cloud** (those fluffy, cotton-like clouds) can weigh over a **million pounds**—even though it looks like it's floating effortlessly!
- Some clouds, like **nimbostratus**, can bring days of gentle, steady rain, while **cumulonimbus** clouds can create strong storms and lightning.
- In places like **Hawaii**, moist air from the ocean hits tall mountains, causing heavy rainfall on one side and a drier climate on the other side—just from a single set of peaks!

Try It Yourself

1. **Cloud Watching:** On a sunny day, lie on your back outdoors and look for different cloud shapes. Can you spot animal shapes, towers, or puffy castles?
2. **Rain Gauge:** Place a wide jar or cup outside during a

rainstorm. Measure how much water you collect in a day.

3. **Water Vapor Experiment:** Take a mirror and breathe on it. You'll see **condensation** form as tiny droplets—just like a mini-cloud you can hold in your hand.

Why It All Matters

Clouds do more than make the sky look pretty. They're **nature's water carriers**, linking lands and seas across the globe. Every raindrop you see, every puddle you splash in, has likely traveled a **long, winding road**—and will travel again. So next time you see a cloud floating by, remember it might be on a **grand adventure**, connecting you to faraway places you've never even seen!

4

Ocean Secrets in Desert Bones

– Seashells Hiding in High Mountains

Imagine standing on a **rocky mountain** or walking through a **dry desert**—the last thing you'd expect to find is a **seashell**! But if you look closely, you just might discover **fossils** of creatures that used to swim in ancient seas. How did these ocean animals end up so far away from the water? The answer is buried deep in **Earth's history**!

How Do Seashells Get Stuck in Rocks?

1. **Ancient Oceans:** Millions of years ago, much of our planet's surface was covered by shallow seas. Areas that are now deserts or mountains used to be completely

underwater!

2. **Sediment and Fossils:** When sea creatures like clams or snails died, their shells sank to the seafloor. Over time, sand and mud piled on top, turning them into **fossils** as the layers hardened into stone.

3. **Earth on the Move:** Over millions of years, **plate tectonics** (shifting sections of Earth's crust) pushed ocean floors upward to form deserts, plains, or mountains. This is why we sometimes find **shell fossils** in places that are super far from any ocean today.

Why Is This Important?

- **Clues to the Past:** Seashell fossils in the desert or on a mountain tell scientists that these places were once under the sea. It's like finding a secret **time machine** buried in stone!

- **Understanding Changes:** When we see these fossils, we learn how Earth's land and water have moved, grown, or shrunk through the ages, helping us predict **future changes**.

- **Treasure Hunt for Knowledge:** Kids and grown-ups alike can become "fossil detectives," using shells in rocks to unravel the story of our planet.

Fun Facts

- **Shark Teeth in the Desert:** Some deserts and dry riverbeds have ancient shark teeth buried in the sand—proof that sharks swam there long ago!
- **Himalayan Surprises:** Even on Mount Everest, the world's tallest mountain, **marine fossils** have been found in the rocky layers near the summit.
- **Dinosaur Seas:** During the time of dinosaurs, giant reptiles also lived in the sea, leaving behind fossils that are sometimes discovered in today's dry, dusty plains.

Try It Yourself

1. **Fossil Field Trip:** If you have a chance to visit a fossil park or a natural history museum, look for **ocean fossils** that were found far inland. Notice their shapes and imagine them as living creatures!

2. **Make a Rock Record:** Collect small rocks (with an adult's permission) and see if you can identify any patterns or layers. Although you might not find a real fossil, you'll start to understand how rocks tell stories.

3. **Draw an Ancient Sea Scene:** Sketch a desert landscape, then draw an underwater world over it. Add fish, coral, and shells to show how the land might have looked millions of years ago!

Why It All Matters

Finding seashells in deserts or mountains is a thrilling reminder that our world is **always changing**. Land that was once a **deep ocean** can become a hot, sandy desert or a towering **mountain range**. Every time you see a fossil from underwater creatures, it's like a **postcard** from Earth's ancient oceans—one more clue in the grand puzzle of our planet's history.

5

The Tree Internet Underground

—Fungi and Roots Talking in the Dark

When you look at a forest, you probably notice the tall trunks, the green leaves, and maybe a squirrel or two. But did you know there's a secret web of life hidden under the ground? It's called the "wood-wide web," and it connects tree roots with thin strands of fungus. Through this underground network, trees can share food, water, and even messages with each other—almost like sending emails in the dirt!

How Does the Wood-Wide Web Work?

1. **Fungal Threads:** Tiny thread-like fungi wrap around or slip inside a tree's roots, forming a special partnership

called a mycorrhiza (my-cor-rye-zuh).

2. **Sharing and Caring:** The fungi help gather nutrients and water from the soil, passing them on to the tree. In return, the tree gives the fungi sugars it makes during photosynthesis.

3. **Tree-to-Tree Talk:** Because the fungus can connect to more than one tree, it acts like an internet cable carrying signals. If one tree is attacked by insects, it might send a warning to other trees so they can make more protective chemicals!

Why Is This Important?

- **Stronger Forests:** Trees that share nutrients and warnings can stay healthier and grow better as a group.
- **Balanced Ecosystem:** A wide variety of fungi and tree species working together creates a stable environment for animals, insects, and even us humans.
- **Nature's Own Network:** This hidden web shows how cooperation helps living things survive, reminding us

that sharing resources can benefit everyone.

Fun Facts

- A single fungus can spread over many acres, making it one of the largest living organisms on Earth!
- Some mushrooms you see above ground are the "fruit" of bigger fungal networks living under the soil.
- Different types of trees can be connected by these fungal threads, forming a grand underground community.

Try It Yourself

1. **Forest Field Trip:** If you visit a forest, peek under fallen leaves or old logs. You might spot thin white threads—those could be the fungus behind the wood-wide web!
2. **Grow Your Own Mushroom:** You can buy a mushroom growing kit (with an adult's help) to watch how fungi grow. Notice the white fuzz that forms before the mushroom appears—those strands are like the ones in the soil.

3. **Draw a Fungus Network:** Sketch a tree and its roots. Then add branching lines of fungus connecting to another tree. Label where nutrients and water might flow!

Why It Matters

The wood-wide web is a reminder that nature works best when living things help each other. Underground fungi and mighty trees prove that even in the darkest dirt, there can be an amazing web of life that keeps forests strong. The more we learn about these hidden connections, the more we see that every part of our planet works together to keep our world healthy and alive.

6

Pollinators in the Big City

—Bees on Skyscrapers and Butterflies on Balconies

Cities might seem like crowded places made of concrete and tall buildings, but if you look closely, you'll find hidden spots where flowers bloom. Rooftop gardens, window boxes, and balcony planters can be lifesavers for tiny creatures like bees and butterflies. These hardworking pollinators zip around collecting pollen and nectar, helping plants grow—even in the middle of a busy downtown!

Pollinators in the City?

1. **Flower Power:** Plants need pollinators to move pollen from one flower to another so they can make seeds. Even

in cities, people plant flowers in parks, rooftops, and balconies, giving bees and butterflies places to feed.

2. **Island Hopping:** Think of gardens like little islands of green in a sea of concrete. Pollinators hop from one patch of flowers to another, helping to keep the city ecosystem alive and colorful.

3. **Healthy Balance:** When pollinators thrive, more flowers grow, which can mean cooler and cleaner air, extra shade, and better habitats for birds and small animals.

Why Is This Important?

- **Food for Everyone:** We rely on pollinators to help produce many fruits and vegetables. If bees and butterflies can survive in the city, we can have fresh produce grown in urban farms and community gardens.

- **Cleaner Air:** Plants that these pollinators visit remove carbon dioxide from the air and give us oxygen in return, making city air healthier to breathe.

- **Nature Connection:** Seeing butterflies and bees reminds us that nature isn't just in forests or fields.

Fun Facts

- Some cities have official programs to build "bee highways," placing flowerbeds along roads and rooftops so bees can easily travel from one patch to the next.
- Monarch butterflies can migrate over 2,000 miles, sometimes passing through large cities, stopping to sip nectar from potted plants on balconies!
- Honey produced by urban bees can taste different from honey in the countryside, because city bees visit a unique mix of flowers.

Try It Yourself

1. **Mini Garden:** If you have a windowsill or balcony, ask an adult to help you plant pollinator-friendly flowers like marigolds, zinnias, or lavender. You might see bees or butterflies dropping by!
2. **Be a Bee Watcher:** Visit a local park on a sunny day and count how many pollinators you spot. Notice their

colors, the flowers they visit, and how they move from plant to plant.

3. **Create a Pollinator Oasis:** Even a small bowl of water with rocks can help thirsty bees and butterflies rest and refuel.

Why It Matters

Pollinators in the big city prove that nature can flourish anywhere. Bees and butterflies play a huge role in keeping our world blooming, and by offering them urban gardens and flower boxes, we're connecting concrete jungles to the green, buzzing world of plants and insects. It's one more way everything on Earth—people, plants, and pollinators—is linked in surprising ways.

7

Volcanoes and Colorful Sunsets

—Fiery Explosions That Paint the Sky

Have you ever gazed at a sunset so dazzling you wanted to capture it forever? Sometimes the sky glows with bright oranges, reds, and pinks. One surprising reason for these brilliant colors can be volcanoes. When a volcano erupts, it sends ash and tiny particles high into the air. If these particles drift far enough, they scatter sunlight in a special way, making sunsets and sunrises extra bold and colorful—even halfway around the world!

How Do Volcanoes Affect the Sky?

1. **Ash and Dust:** During an eruption, the volcano shoots

ash, dust, and gases into the atmosphere.

2. **Sunlight Show:** Tiny particles spread across the sky, bending and filtering the sunlight that passes through them. This can turn an ordinary sunset into a glowing masterpiece of color.

3. **Traveling Clouds:** Winds can carry volcanic ash thousands of miles, so you might see spectacular skies far from where the volcano erupted!

Why Is This Important?

- **Global Impact:** Volcanic ash can circle the planet, affecting air travel, weather patterns, and even global temperature.

- **Clues for Scientists:** Studying how volcanic particles move helps scientists learn more about Earth's atmosphere and how natural events can change it.

- **Art of Nature:** These vivid sunsets remind us that natural processes—like volcanoes—can create beauty along with destruction.

Fun Facts

- The 1883 eruption of Krakatoa in Indonesia was so big that people around the world saw strange, colorful skies for months afterward.
- Volcanic ash can reflect sunlight away from Earth's surface, cooling the planet slightly after a massive eruption.
- Not all volcanic eruptions are explosive—some volcanoes ooze lava slowly, releasing fewer particles into the air, which might not change sunsets much.

Try It Yourself

1. **Sunset Watch:** On days after a volcanic eruption is in the news, look carefully at the sky during sunset. You may notice richer reds or pinks if the ash drifts your way.
2. **Make a Sunset Jar:** Fill a clear jar with water and a little bit of milk (or another cloudy liquid). Shine a flashlight through it in a dim room. You'll see how light scatters, creating sunset-like colors.

Why It Matters

Volcanoes don't just shape the land; they also shape the sky. Even if an eruption happens on the other side of the planet, its effects can travel and color our world in surprising ways. By learning about these fiery mountains, we discover yet another amazing connection between Earth's surface and the beautiful skies above us!

8

Ants, Aphids, and Secret Farming

—*How Tiny Bugs Work Together*

When most people think about farmers, they imagine fields of corn or rows of vegetables. But did you know that tiny ants are like mini farmers too? In many gardens and forests, ants "farm" small insects called aphids, protecting them and helping them grow. In return, the aphids give the ants a sweet treat. It's a clever partnership hidden right under our noses!

How Does This Relationship Work?

1. **Aphid Juice:** Aphids suck sap from plant stems and leaves. They turn that sap into a sugary liquid called honeydew.

2. **Ant Bodyguards:** Ants love honeydew and keep aphids safe from predators like ladybugs. In exchange, the ants get to feast on the sticky sweetness.

3. **Plant Impact:** Sometimes too many aphids can weaken the plants they feed on, but ants will move them around—almost like ranchers herding cows—so the aphids can find fresh spots for sap.

Why Is This Important?

- **Ecosystem Roles:** Aphids, ants, and plants form a three-way connection. Observing them helps scientists understand how small insects affect bigger ecosystems.

- **Balance of Nature:** Ants keep aphid enemies away, which can be good or bad depending on whether we want fewer aphids on garden plants.

- **Big Lessons from Little Bugs:** Seeing ants and aphids cooperate reminds us that surprising partnerships can help all sides survive.

Fun Facts

- Some ant species gently tap aphids with their antennae to "milk" them for drops of honeydew.
- Aphids come in different colors—green, black, or red—and often group together on plant stems.
- Ants sometimes fight off ladybugs to protect their "herds" of aphids.

Try It Yourself

1. **Garden Safari:** If you have a garden or visit a park, check under leaves or on plant stems. You might see aphids and the ants looking after them!
2. **Draw a Bug Farm:** Sketch an ant, a group of aphids, and a plant leaf. Label where the honeydew might come from and show how ants "milk" aphids.
3. **Observe Changes:** Check on the same plant every day to see if there are more aphids, fewer ants, or any ladybugs stopping by.

Why It Matters

Ants and aphids may be small, but their teamwork shows how nature finds creative ways for living things to help one another—sometimes in odd or unexpected ways. By paying attention to the tiny farms in our own backyards, we discover that cooperation is an important part of our world, no matter how big or small we are!

9

Bird Poop That Grows Coral

—Seabirds Feeding Reefs with Droppings

Coral reefs are often called the "rainforests of the sea" because they're full of bright colors and all kinds of ocean life. But did you know that part of what helps these reefs thrive comes from above the water? Seabirds that nest on islands leave droppings—often called guano—that wash into the ocean, providing nutrients to feed coral. It might sound a little gross, but it's a powerful way birds help keep coral reefs healthy!

How Do Seabirds Help Corals Grow?

1. **Nutrient Boost:** Bird poop is rich in nitrogen and phosphorus, important nutrients for ocean plants and

algae. Corals rely on algae that live inside them, so extra nutrients mean stronger, healthier reefs.

2. **From Land to Sea:** When it rains, bird droppings on coastal cliffs or island shores can flow into the ocean, carrying nutrients to nearby coral reefs.

3. **Island-to-Reef Connection:** Many coral reefs form around small tropical islands. Seabirds nesting there create a steady supply of nutrients, linking land and sea in a surprising way.

Why Is This Important?

- **Strong Coral Reefs:** Healthy reefs support countless fish, crabs, sea turtles, and other creatures. Bird droppings might seem unimportant, but they boost coral growth and keep reefs alive.

- **Food for Fish:** Algae that grow from these nutrients become food for fish and other reef animals, which then become food for bigger predators like sharks or dolphins.

- **Protecting Coasts:** Coral reefs help shield shorelines from strong waves. By keeping reefs healthy, seabirds indirectly protect both people and wildlife on land.

Fun Facts

- In some places, bird droppings—called guano—are collected and sold as fertilizer for farms because it's so nutrient-rich.
- Seabirds like boobies, frigatebirds, and gannets often nest in huge colonies, creating massive piles of droppings over time.
- Coral reefs cover less than 1% of the ocean floor but support about 25% of all marine species, making them one of Earth's most diverse habitats.

Try It Yourself

1. **Coral Reef Map:** Find a map of tropical islands that have coral reefs around them. Imagine how rain might carry nutrients from bird colonies into the water.
2. **Local Birds:** Research seabirds that live near coral reefs.

What do they eat, and how could that affect the nutrients in their droppings?

3. **Food Web Drawing:** Sketch a simple food chain linking bird droppings to algae, coral, fish, and bigger predators. Notice how each step needs the one before it.

Why It All Matters

Bird poop might seem like a messy nuisance, but in tropical island ecosystems, it's a treasure trove of nutrients for coral reefs. This shows how land and sea are connected in fascinating ways—even through something as unexpected as droppings. By understanding these links, we learn that every piece of the puzzle, big or small, helps keep our planet healthy and alive.

10

Giant Turtles and Ocean Currents - Adventurers of the Sea!

—Swimming Paths That Circle the Globe

Sea turtles are like adventurous travelers of the ocean. From the moment they hatch on sandy shores, these gentle giants can journey thousands of miles across Earth's waters, riding along the powerful "highways" we call ocean currents. It's a bit like hopping on a moving walkway that never stops! Even though the sea can seem endless, turtles use these currents to connect coasts and marine life all over the world.

How Do Ocean Currents Help Turtles Travel?

1. **Natural Conveyer Belts:** Ocean currents flow through the seas in giant loops, pushed by wind, temperature changes, and Earth's rotation. Turtles catch these currents to glide along with much less effort.

2. **Navigation Magic:** Scientists think turtles use Earth's magnetic fields and the angle of the Sun to figure out where they are and which way to go. Combined with currents, it's like having a built-in map and a strong tailwind!

3. **Food and Safety:** By following currents, turtles find areas rich in jellyfish, seaweed, or crabs to eat. They can also avoid some predators by sticking to well-known routes.

Why Is This Important?

- **Linking Distant Shores:** A turtle hatched on one beach might feed near coral reefs far away and then return home years later to lay eggs, tying different coasts and marine habitats together.

- **Healthy Oceans:** Sea turtles eat jellyfish, sea grasses, and algae, helping keep those populations balanced. Without turtles, some creatures can overgrow, disturbing the ocean's natural harmony.

- **Indicator Species:** If sea turtle populations start to shrink, it often signals pollution, overfishing, or harmful changes in the environment. Protecting turtles can also help many other sea creatures.

Fun Facts

- **Marathon Swimmers:** Leatherback sea turtles can cross entire oceans, swimming over 10,000 miles in just one year.

- **Built-in Compass:** Baby turtles use moonlight to find the ocean when they hatch. As adults, they use the Earth's magnetic field like a compass to navigate.

- **Longevity:** Some sea turtles can live 50 years or more, giving them plenty of time for epic ocean adventures.

Try It Yourself

1. **Map Their Path:** Look up migration routes of sea turtles online (with an adult's help). Notice how many countries they pass or how many oceans they cross.
2. **Turtle Tracker:** Draw a world map on paper and mark spots where different turtle species nest, feed, or travel. Imagine how they use currents to get around.

Why It Matters

Sea turtles remind us that oceans aren't just random bodies of water. They're connected by currents that carry animals, nutrients, and even drifting seeds around the globe. By learning how turtles rely on these powerful flows, we see how a single creature's journey can link entire continents—further proof that our planet's waters, lands, and living beings are all wonderfully interconnected.

11

The Desert Frogs' Rain Dance

—Buried Sleepers That Predict Storms

Imagine living in a place where rain is scarce, and the land is hot and dry most of the time. That's the challenge desert frogs face! Instead of hopping around all day, many of these clever amphibians bury themselves underground. They wait there—sometimes for months—until the sky finally opens up. When rain soaks the desert, these frogs pop up, sing loudly, and lay their eggs in puddles that might last only a few days.

How Do They Survive So Long Without Rain?

1. **Burrow Masters:** Desert frogs, like spadefoot toads, use strong back legs to dig deep in the sand or soil.

2. **Moisture Lock-In:** While underground, they slow down their body's activities to save energy and water.

3. **Time to Shine:** When a storm arrives, water seeps into the soil, waking the frogs. They rush above ground to eat, mate, and lay eggs before the puddles dry up!

Why Is This Important?

- **Nature's Alarm:** People who live near deserts say they can guess a coming rainstorm by hearing frog calls at night.

- **Rapid Life Cycle:** Desert frog eggs often hatch within a day or two, and tadpoles grow very quickly to beat the desert heat.

- **Ecosystem Boost:** When frogs are active, they become food for birds and other animals, which helps balance the entire desert community.

Fun Facts

- **Chorus of Croaks:** After a good rain, the nighttime desert can sound like a massive frog orchestra performing in the dark!

- **Surprise Discovery:** Early explorers didn't know frogs could survive underground, so they were shocked to see frogs "appear" out of nowhere after a rainstorm.

- **Adaptable Skin:** Many desert frogs have skin that helps seal in moisture, letting them wait out dry spells for months.

Try It Yourself

1. **Frog Call Search:** Listen to recordings of desert frog calls online (with an adult's help). Notice how each species has a unique sound.

2. **Frog Artwork:** Draw a desert frog burrowing underground. Show how it might stay moist and safe during dry weather.

3. **Local Amphibians:** If you don't live in a desert, look up frogs or toads in your area. Do they have special ways of handling heat or cold?

Why It Matters

Desert frogs show us that life can flourish in the toughest places with the right survival tricks. Their hidden burrows and joyful rain dances remind us that Earth's creatures are full of surprises. By understanding how these frogs adapt to extreme conditions, we see again how living things—even in deserts—are linked to the rain, the land, and each other in countless unexpected ways.

12

Whale Farts and Climate Change

—How Gases from the Deep Affect the Air We Breathe

Did you know that giant whales in the ocean can release huge bubbles of gas? Some people jokingly call them "whale farts," but this unusual fact points to something bigger: everything happening in the ocean can affect the air above us. From plankton blooms to whale digestion, the ocean is linked to the Earth's climate in surprising ways!

What's the Connection Between Whales and Climate?

1. **Plankton Partners:** Whales eat tiny ocean creatures

called plankton. When whales poop, they add nutrients back into the water, helping plankton grow. Plankton absorb carbon dioxide (CO_2)—a gas that warms our planet.

2. **Gas Release:** Like all mammals, whales produce gases in their bodies. These gases can bubble to the surface. Though whale farts aren't the main factor affecting climate, they're a funny reminder that living creatures release different gases into the air and water every day.

3. **Carbon Sinks:** The deep ocean stores huge amounts of carbon. When whales dive and feed, they help move carbon between surface waters and deeper areas, influencing Earth's climate over time.

Why Is This Important?

- **Ocean Health:** More plankton means more food for fish, birds, and other sea life, creating a healthier ocean that also absorbs more CO_2.

- **Climate Balance:** Since having too much CO_2 in the air can cause the planet to heat up, strong whale populations can indirectly help keep Earth cooler.

- **Learning from Nature:** Watching how whales interact with the ocean helps scientists see how gases move around and affect our world.

Fun Facts

- Blue whales are the largest animals on Earth—bigger than any dinosaur that ever lived—and they can create some enormous bubbles of gas!
- Whale poop is sometimes bright orange or pink, depending on the tiny sea creatures the whales have eaten.
- Some whales migrate thousands of miles, spreading nutrients over many different parts of the ocean.

Try It Yourself

1. **Ocean Food Web:** Draw a diagram with whales at the top and plankton at the bottom. Show how nutrients and carbon move between them.
2. **Carbon Chat:** Research (with an adult) what carbon

dioxide does in our atmosphere and why balancing it helps keep temperatures just right.

3. **Whale Watching:** If you ever get the chance, go on a whale-watching trip (with an adult). Observe how big they are and think about how they might affect the entire ocean around them.

Why It Matters

Even something as odd as a whale fart can remind us how Earth's oceans and atmosphere work together. Whales help spread nutrients, feed plankton, and keep our oceans healthy. When the ocean is healthy, it can store more carbon, which helps balance the climate on land. By protecting whales and their habitats, we also protect the air we breathe—another example of our planet being a web of amazing, interconnected life.

13

Squirrels That Plant Forests

—Hiding Nuts That Grow into Tall Trees

Have you ever watched a squirrel scurry around your yard, burying acorns or seeds in the ground? These busy animals aren't just storing food for winter—they're also accidentally planting new forests! Squirrels sometimes forget where they've hidden their nuts, leaving those seeds to sprout into baby trees. Over time, those young trees can grow into mighty forests that benefit countless animals (including people).

How Do Squirrels Help Forests Grow?

1. **Acorn Collectors:** Squirrels gather acorns, chestnuts, and other seeds, then dig tiny holes to hide them away.

2. **Forgetting Spots:** Not every seed is eaten. Some remain buried in the soil, waiting for the right conditions to sprout.

3. **Natural Gardeners:** As these seeds grow, new trees appear in spots where they might not have grown otherwise, helping forests spread and stay healthy.

Why Is This Important?

- **New Habitats:** Each new tree can become a home for birds, insects, and other wildlife.
- **Food Chain Boost:** Forgotten seeds turn into trees that produce more nuts, feeding future generations of squirrels and other creatures.
- **Erosion Control:** Tree roots help hold soil in place, preventing heavy rain from washing it away and protecting the land from erosion.

Fun Facts

- A single squirrel can bury hundreds of nuts each year. Imagine how many new trees that could create over time!

- Squirrels sometimes move nuts multiple times, spreading seeds across larger areas of the forest.
- Some trees, like oaks, rely on animals like squirrels to disperse their acorns because the seeds can't move on their own.

Try It Yourself

1. **Forest Scavenger Hunt:** The next time you visit a park or wooded area, look for small holes or disturbed spots in the soil—these could be squirrel hiding spots!

2. **Plant a Seed:** With an adult, try planting an acorn or nut in a pot. Watch how it grows and imagine the difference squirrels make in the wild.

3. **Sketch a Squirrel Scene:** Draw a squirrel burying acorns. Label where seeds might be planted and what kind of tree could sprout.

Why It Matters

Squirrels seem like simple backyard visitors, but their nut-burying habits have major effects on forests. By forgetting some of the seeds they hide, these furry "gardeners" plant the next generation of towering trees. It's a reminder that small everyday actions in nature—like a squirrel hiding a snack—can lead to big, beautiful changes in our environment!

14

THE MIGHTY MICROSCOPIC MITES

—Tiny Helpers Living on Your Eyelashes

Have you ever felt a tickle on your eyelashes? It might be from tiny creatures called mites—so small you can't see them with your eyes alone. These microscopic mites are like mini cleaning helpers living on the hairs around your eyes. While it might sound strange or even gross, they actually do a useful job by munching on oils and dead skin cells that build up on your eyelids.

HOW DO EYELASH MITES LIVE AND WORK?

1. **Home Sweet Home:** Mites make their homes in the pores and hair follicles around your eyes, where they

find plenty of oils and dead skin cells to eat.

2. **Tiny Cleaners:** By feeding on these oily bits, mites help keep your lashes and eyelids from getting too greasy. Think of them as a mini clean-up crew!

3. **Out of Sight:** Because they're so small (less than half a millimeter), you don't notice them going about their business. They're nearly invisible to us without a microscope.

Why Is This Important?

- **Balanced Skin:** Having these mites around can help manage the natural oils on your eyelids, preventing too much buildup.
- **Living Ecosystem:** Even our bodies host tiny living creatures. Mites are part of the ecosystem that makes up our skin.
- **Cool Science:** Studying eyelash mites teaches scientists about how humans and microscopic organisms can help each other without us ever noticing.

Fun Facts

- Mites are related to spiders and ticks, but eyelash mites are usually harmless to people.
- Some people have more mites than others, depending on their body chemistry.
- Mites mostly come out at night to feed, so you're asleep and never feel them crawling!

Try It Yourself

1. **Learn About Microbes:** Research (with an adult) all the types of tiny life on human skin—bacteria, fungi, and mites—and discover just how crowded your skin really is.
2. **Draw a Mite:** Sketch a super-enlarged version of an eyelash mite. They have teeny legs and a long, tube-shaped body.

Why It Matters

Eyelash mites might seem icky, but they show how even the smallest creatures can be helpful. They're part of a hidden community that keeps our skin healthy and balanced. By understanding these "invisible roommates," we uncover another example of how every living thing—no matter how tiny—plays a role in the grand design of nature.

15

Fireflies and City Lights

—Flickering Bugs Searching for Darkness

Have you ever seen a soft green or yellow glow floating through the night? That's probably a firefly, also called a lightning bug. Fireflies flash their lights to talk to each other and attract mates. But in big cities, bright streetlights and signs can make it hard for them to communicate. This happens because artificial lights create light pollution, washing out the fireflies' glowing signals.

Why Do Fireflies Need Darkness?

Fireflies use bioluminescence—light produced inside their bodies—to send messages. Males often flash a pattern to impress females, and females answer with their own glow. When there's too much artificial light, fireflies might struggle to see each other's signals. This can mean fewer chances to meet, mate, and lay eggs, putting future firefly generations at risk.

Why Is This Important?

- **Nature's Nighttime Show:** Fireflies bring magic to summer evenings, and losing them means losing a special part of outdoor life.
- **Pollinator Partners:** Some types of fireflies also help pollinate flowers, though it's not their main job. Every bit of pollination can help plants thrive.
- **Healthy Ecosystem:** Fireflies need specific habitats—damp areas, tall grass, and darkness. Protecting these habitats often means helping many other night-loving creatures, too.

Fun Facts

- Each firefly species has its own unique flashing pattern, like a secret code only they understand.

- Firefly "light" doesn't give off heat, so it won't burn you if you hold one.

- In some parts of the world, there are "synchronous fireflies" that light up in perfect rhythm, creating a breathtaking glow.

Try It Yourself

1. **Lights Out:** If possible, turn off outdoor lights or close curtains at night so fireflies can find the darkness they need.

2. **Firefly Watch:** Look for blinking lights on warm summer nights. Notice if certain flashes seem faster, slower, or in groups—these might be different firefly species.

Why It All Matters

Fireflies and their glowing lights remind us how important darkness can be for certain creatures. As more city lights shine through the night, fireflies—and many other nocturnal animals—have to work harder to survive. By preserving some darkness in our neighborhoods, we help these flickering bugs keep flashing their magical signals.

16

Dolphin Conversations in the Sea

—Clicking and Whistling Across Oceans

Have you ever wondered how dolphins find each other or navigate the wide, open ocean? These playful marine mammals communicate using clicks, whistles, and squeaks. Each sound carries messages that can travel far underwater, helping dolphins bond with friends, find food, and stay safe from dangers like sharks. In a way, it's like a secret language no human can fully speak!

How Do Dolphins Talk?

1. **Clicking for Echolocation:** Dolphins make quick clicking sounds and listen for the echoes that bounce off objects. This tells them how far away things are, acting like a built-in sonar.

2. **Whistles and Squeaks:** Dolphins also create whistles and squeaky sounds to call to each other. Scientists believe each dolphin may even have a unique "signature whistle," like a personal name.

3. **Body Language:** Besides sounds, dolphins communicate by leaping out of the water, slapping their tails, and swimming in specific patterns. It's not just what they say, but how they move!

Why Is This Important?

- **Ocean Navigation:** Echolocation allows dolphins to hunt fish and avoid obstacles, even in murky or dark water.

- **Social Bonds:** Dolphins live in groups called pods, and their clicking and whistling keeps them united, almost like a team huddle.

- **Scientific Clues:** Studying dolphin sounds helps scientists learn more about ocean health. When dolphins have trouble communicating, it might mean pollution or noise from ships is disrupting their world.

Fun Facts

- Dolphins are part of the cetacean family, which also includes whales and porpoises.
- Some dolphins can leap up to 20 feet out of the water, making a big splash and showing off for their pod.
- The loud clicks used for echolocation can reach over 200 decibels—a powerful signal in the underwater realm!

Try It Yourself

1. **Echo Test:** Clap your hands loudly in a big empty room or a canyon. Listen for the echo. Imagine a dolphin using those sound bounces underwater!

2. **Dolphin Dictionary:** Look up different dolphin sounds online (with an adult). Notice how whistles, clicks, and squeaks all differ.

Why It Matters

Dolphins remind us that the ocean is brimming with communication networks we're only beginning to understand. Their chats, clicks, and whistles help them stay connected and survive in vast waters.

17

Caves That Make Cheese

—Tasty Secrets from Deep Underground

Have you ever wondered how some cheeses get their bold, unique flavors? In certain parts of the world, people age cheese in **caves**, where cool, damp conditions help special molds grow. These molds give each cheese a distinct taste, texture, and smell—making it truly unforgettable. It's like nature's underground kitchen, where cheese becomes a delicious work of art!

How Do Caves Help Make Cheese?

1. **Perfect Climate:** Caves are naturally cool and humid, which slows down how fast cheese ripens. This lets flavors develop more deeply over time.

2. **Friendly Molds:** Certain molds that thrive in caves attach themselves to the cheese rind. These molds can give cheese a soft, creamy inside or a sharp, tangy taste.

3. **Slow and Steady:** Since caves aren't too hot or too cold, cheese can age slowly, letting good bacteria and molds work their magic without spoiling the cheese.

Why Is This Important?

- **Rich Flavors:** Cave-aged cheeses are famous for intense flavors you can't get from regular refrigeration.

- **Cultural Tradition:** People in countries like France, Italy, and Switzerland have used caves to age cheese for centuries, passing down methods from one generation to the next.

- **Microbial Diversity:** Studying the molds and bacteria in cave cheeses teaches scientists about beneficial microbes that can help us produce tastier, healthier foods.

Fun Facts

- Some caves where cheese is aged have been used for

hundreds of years, with certain families running the same cheese-aging process for generations.

- Roquefort cheese, a famous French variety, gets its blue veins from molds that only grow in limestone caves near the village of Roquefort.
- Not all cheeses need a cave—some are aged in temperature-controlled rooms built to act like artificial caves!

Try It Yourself

1. **Taste Test:** Visit a local cheese shop or grocery store (with an adult) and try a cave-aged cheese. Notice how it compares to regular cheeses you already know.
2. **Explore Food Origins:** Look up different cheeses from around the world. Which ones are aged in caves, and what makes each special?
3. **Make a Cheese Poster:** Draw or print pictures of different cheese varieties and label where they're from, how they're aged, and what they taste like.

Why It Matters

From the surface, caves might just seem like dark, damp places, but inside, they're home to age-old cheese-making traditions. This is another example of how natural environments—even deep underground—shape the food we eat and the cultures we celebrate. It's proof that **Earth's hidden spaces** can be some of our most valuable resources, bringing unique flavors and fascinating stories straight to our plates!

18

Pollen from Africa in America

—*One Flower's Journey Across Oceans*

Have you ever wondered how a flower from one continent could end up influencing plants and people on another continent far away? Tiny pollen grains can float on the wind for thousands of miles, crossing entire oceans before settling down in new places. It's like a worldwide air-travel system for plants—one that helps shape ecosystems, even across continents!

How Does Pollen Travel So Far?

1. **Wind Power:** Pollen grains are incredibly light and can be picked up by strong winds in places like the Sahara

Desert. Once caught in air currents, they may drift high above Earth's surface.

2. **Long Flights:** Some winds blow for hundreds or thousands of miles across oceans, carrying pollen with them. Eventually, these grains land on another continent—or even another hemisphere!

3. **Settling Down:** While most pollen grains won't grow into new plants unless they land on the right kind of flower, they still bring tiny nutrients or can even change the air quality in small ways.

Why Is This Important?

- **Plant Diversity:** Traveling pollen can help spread genes between different plant populations, boosting the variety of flowers and crops.

- **Allergy Clues:** Sometimes, pollen from faraway places can trigger sneezing and watery eyes in people who live nowhere near the original plants!

- **Scientific Detective Work:** Finding African pollen

in South America, for example, proves that winds can connect distant ecosystems in surprising ways.

Fun Facts

- Scientists studying ice cores in places like Greenland or Antarctica often find pollen grains from distant lands, showing how far pollen can roam.
- Dust storms in the Sahara can carry not just pollen but also nutrients that help fertilize rainforests in the Amazon.
- Some plants have specially adapted pollen that's super light and floaty, meant to travel far on the breeze.

Try It Yourself

1. **Pollen Watch:** In spring or early summer, look at the dust on your windowsill or car. Look at it under a microscope with the help of an adult. Some of that powdery stuff could be pollen from plants near and far!
2. **Map the Winds:** With an adult, explore global wind patterns online. Notice how some major wind belts go

right across oceans—perfect for long-distance pollen travel.

3. **Nature Journal:** Keep track of when you start sneezing or see lots of pollen dust. Compare notes each day to see if windy weather makes a difference.

Why It Matters

Tiny pollen grains show us how nature can bridge huge distances, linking deserts and rainforests, mountains and coastlines. Even something as small as pollen can have a big impact, affecting plant life, weather patterns, and even our own health. It's yet another reminder that the whole planet—people, plants, and places—is connected by invisible threads in the air all around us!

19

Frogs and Medicine in the Rainforest

—*How Nature's Potions Heal People*

Deep in tropical rainforests, there are brilliantly colored frogs that carry powerful chemicals in their skin. These "poison frogs" may look cute, but some of them can pack a serious punch! While their toxins keep predators away, scientists have discovered that some of these chemicals can also be used in medicines to treat pain, infections, or even serious diseases. It's nature's pharmacy at work!

How Do Rainforest Frogs Help Medicine?

1. **Colorful Warnings:** Frogs like the poison dart frog flash bright reds, blues, and yellows to warn predators: "Don't eat me—I'm toxic!"

2. **Skin Chemistry:** These frogs produce special toxins on their skin. In small, controlled amounts, certain toxins can help doctors create new medicines.

3. **Lab Discoveries:** Scientists study frog venom in laboratories to see if it can fight bacteria or block pain signals, which might lead to better treatments for people.

Why Is This Important?

- **New Cures:** The rainforest holds an incredible variety of plants and animals with unique chemicals. Studying them could lead to breakthroughs in medicine.

- **Protecting Habitats:** If rainforests are destroyed, we lose not only rare frog species but also the chance to discover lifesaving compounds.

- **Teamwork with Nature:** Learning how frogs' bodies make and use toxins teaches us about the balance between living things and their environments.

Fun Facts

- **Poison Dart Name:** Indigenous people once used secretions from these frogs on their blow darts for hunting. The toxins were strong enough to paralyze animals!

- **Brightest of All:** The golden poison frog is considered one of the most toxic creatures on Earth—just touching it can be dangerous!

- **Frog Cafeteria:** Scientists believe some frogs get their toxins by eating special insects or plants in the rainforest, which is why frogs raised in captivity often have weaker poisons.

Try It Yourself

1. **Rainforest Research:** Look up (with an adult) different rainforest frogs. Notice the range of colors and patterns that warn off predators.

2. **Draw a Poison Frog:** Pick your favorite poison dart frog and create a brightly colored illustration. Label the spots or stripes that make it stand out.

Why It All Matters

Frogs and their remarkable skin chemicals remind us that the natural world is filled with hidden gifts waiting to be discovered. By protecting rainforests, we protect not only these vibrant amphibians but also the possible medicines they carry. It's one more example of how wildlife and human health are connected—sometimes in the most surprising ways!

20

Bats That Pollinate Bananas

—Night Flyers That Make Your Breakfast Possible

When the sun goes down and the world seems quiet, bats spring into action. While many people think of bats as spooky creatures, they actually do an important job in nature—**pollinating** plants, including some that give us delicious fruits like bananas! By flapping from flower to flower under the moonlight, bats help these plants produce the tasty treats we enjoy.

How Do Bats Pollinate Plants?

1. **Nighttime Gathering:** While bees and butterflies usually pollinate during the day, bats prefer the dark.

They fly around at night, visiting special flowers that open under moonlight.

2. **Flower Dust:** As a bat sips nectar, pollen sticks to its fur. When it moves to the next flower, some of that pollen brushes off, helping the plant make seeds and fruit.

3. **Banana Boost:** Certain banana varieties depend on bats to move pollen between flowers. Without bats, these banana plants could struggle to grow fruit.

Why Is This Important?

- **Midnight Pollination:** Night-pollinating bats make sure flowers that bloom in the evening still get what they need to produce fruit.

- **Fruit for Breakfast:** Many of our favorite foods—like bananas, mangoes, and guavas—rely on bats for pollination.

- **Ecosystem Helpers:** By pollinating plants, bats encourage new seeds to grow into trees and shrubs that provide homes for other animals.

Fun Facts

- Some flowers produce a strong scent at night to attract bats and moths, like a secret invitation to a moonlit buffet!
- The tube-lipped nectar bat of South America has a tongue longer than its entire body—perfect for reaching deep into flowers.
- Bats are key pollinators in tropical regions, ensuring we have a steady supply of fruits, nuts, and seeds.

Try It Yourself

1. **Night Nature Walk:** (With an adult) go outside at dusk or after dark and observe if bats are flying around. Notice how fast they swoop!
2. **Banana Investigation:** Look at a banana flower photo or video online. Discuss with friends or family how bats might gather nectar from the flower.

3. **Pollinator Comparison:** Make a chart showing daytime pollinators (bees, butterflies) and nighttime pollinators (bats, moths). Compare how they work differently but achieve the same goal!

Why It Matters

Bats might get a spooky reputation, but they're truly superheroes when it comes to pollination. Their nighttime visits keep certain plants thriving, helping to produce fruits that land on our tables every day. It's one more way our world connects in unexpected ways—where even a bat's midnight snack can lead to a sweet banana for your breakfast!

21

Crabs That Clean Beaches

—*Sandy Shore Keepers Hard at Work*

Next time you visit the beach, you might spot little creatures scuttling around in the sand. These are **crabs**, and they do much more than just pinch and hide in shells! Some crabs help keep beaches clean by eating rotting plants, dead fish, and other leftover scraps washed up by the waves. By tidying up the shore, they play a huge role in keeping the beach healthy for both wildlife and people.

How Do Crabs Clean Up the Beach?

1. **Scavenger Snacks:** Crabs are natural scavengers, meaning they eat dead or decaying materials. This helps

remove smelly waste before it can rot and attract flies.

2. **Soil Mixers:** While digging their burrows in the sand, crabs mix nutrients into the upper layers of the beach, creating better conditions for small plants or other creatures.

3. **Food Source for Others:** Crabs themselves become meals for birds, fish, and other animals, forming a link in the coastal food web.

Why Is This Important?

- **Healthy Shorelines:** When crabs remove decaying matter, the beach stays cleaner and smells better, making it pleasant for everyone.

- **Balancing Act:** By burrowing in the sand, crabs also help keep the ecosystem balanced, allowing other organisms to thrive.

- **Protecting Coastal Life:** Clean beaches mean safer homes for sea turtles and seabirds that nest or feed along the shore.

Fun Facts

- Fiddler crabs get their name from their one oversized claw, which looks a bit like a violin's bow when they wave it to attract mates.
- Ghost crabs are quick runners and can disappear into their burrows in the blink of an eye!
- Some hermit crabs carry seashells on their backs for protection, swapping shells as they grow bigger.

Try It Yourself

1. **Beach Safari:** On your next trip to the shore, watch for crab tracks or small holes in the sand. Gently scoop up a handful of sand to see if any tiny crabs scurry out!
2. **Crab Sketch:** Draw a crab species you find interesting—like a ghost crab or a fiddler crab. Label its claws, legs, and shell.

Why It Matters

Crabs may be small, but their impact on beach health is mighty. By gobbling up leftovers and tunneling through the sand, they keep shorelines clean and thriving for countless other creatures. So the next time you see a crab scuttling across the sand, remember it's not just a funny beach critter—it's part of a grand system that helps protect and balance the entire coastal environment!

22

Super Snail Slime

— Trails That Lead to New Adventures

Have you ever spotted a shiny, slimy trail zigzagging across a garden path? That glistening line is **snail slime**, and it's a big part of how snails get around. They may seem slow, but they're masters at creating their own slippery highway! Their mucus-like slime helps them climb walls, explore new places, and even share information with other snails.

How Do Snails Use Their Slime?

1. **Smooth Travels:** The slime acts like a cushion under a snail's body, allowing it to glide across rough surfaces—or even straight up a wall.

2. **Snail GPS:** Other snails can follow these trails to find

food or potential mates, almost like a map someone else left behind.

3. **Protection:** The mucus helps keep a snail's soft body from drying out, which is super important on hot days. It can also make it harder for predators to grip them.

Why Is This Important?

- **Garden Health:** By exploring gardens, snails help recycle decaying plant material, returning nutrients to the soil.
- **Food Web Role:** Snails are a food source for birds, beetles, and other animals, linking them to a larger ecosystem.
- **Nature's Clues:** Studying snail slime and behavior gives scientists clues about environmental conditions, like moisture levels and soil health.

Fun Facts

- Some land snails can have more than **14,000 tiny teeth** arranged on their tongue-like organ, called a radula, which they use to scrape up plants.

- Snail slime is sometimes used in skin care products because it contains proteins and vitamins that may help moisturize skin.

- Certain snails have spiral-shaped shells that twist either to the left or to the right—no two shells are exactly the same!

Try It Yourself

1. **Slime Hunt:** If you have a garden, look for silvery trails on leaves or pavement in the morning or after rain. See if you can trace where the snail traveled.

2. **Mini Habitat:** Place a snail in a small terrarium (with an adult's help) and offer fresh greens. Observe how it moves and leaves behind slimy paths!

3. **Snail Sketch:** Draw a snail and label its shell, body, and the slime trail. Imagine what it might see or smell as it glides along.

Why It Matters

Snail slime might look gooey and strange, but it's an amazing adaptation that helps these slow-moving creatures thrive. By leaving trails for others to follow and helping recycle plant matter, snails show us that even the smallest, slimiest parts of nature can play a big role in keeping our world connected and healthy!

23

Lightning and Mushrooms

—Storm Sparks Supersize Shrooms!

Have you ever wondered what happens deep in the forest when lightning strikes? Besides dazzling thunder and rain, there's another surprising effect: lightning can **supercharge** mushrooms! Certain mushrooms grow faster and bigger after being zapped by tiny electric jolts similar to lightning. It's as if these forest fungi get an energy boost from the sky.

How Does Lightning Affect Mushrooms?

1. **Electric Encouragement:** Scientists have found that mushrooms given small electric shocks can sprout more quickly. It's similar to how plants respond to lightning by growing better in the days following a storm.

2. **Nutrient Release:** Lightning can help break down nutrients in the soil, making them easier for mushrooms to absorb.

3. **Natural Boost:** Although we don't see the process up close, the results can be huge mushrooms popping up suddenly after a big thunderstorm!

Why Is This Important?

- **Food in the Forest:** Mushrooms feed many creatures, from tiny insects to larger animals like deer. More mushrooms can mean more food for everyone.

- **Soil Health:** Fungi help break down dead plants and logs, returning nutrients to the earth. When they're plentiful, forests stay strong and healthy.

- **Scientific Curiosity:** Learning how lightning interacts with fungi can inspire new ideas in agriculture and environmental science.

Fun Facts

- Some types of mushrooms can appear almost

overnight, growing inches in a single day, especially after a storm.

- Mushroom "fairy rings" are circles of mushrooms that form as the fungal network under the soil expands outward.
- Certain glow-in-the-dark mushrooms exist, which seem extra magical when lightning lights up the sky!

Try It Yourself

1. **Storm Watch:** After a thunderstorm, head outside (with an adult) to see if you can spot new mushrooms popping up in your yard or a nearby park.

2. **Nature Log:** Keep track of how many days pass between a storm and when you see mushrooms. Note their colors, shapes, and how quickly they grow.

3. **Fungi Exploration:** Research (with an adult) different mushroom species. Find out which ones are safe to eat, but remember—never taste a wild mushroom unless an expert says it's okay!

Why It Matters

Lightning isn't just a scary flash in the sky; it can have a surprisingly positive effect on forests. By helping mushrooms grow, storms support the entire ecosystem, from the soil to the animals that depend on this tasty resource. It's a shocking (but true) reminder that nature's energy can spark life in unexpected ways!

24

The Solar Flare Surprise

—Sun Storms That Switch Off Your Lights

Did you know that the Sun doesn't just shine—it also throws giant tantrums called **solar flares**? These fiery explosions on the Sun's surface send out waves of charged particles and energy that can travel millions of miles to reach Earth. When they arrive, they can interfere with our power grids, satellites, and even GPS signals, causing blackouts or strange glitches. It's a cosmic reminder of how connected we are to our star!

How Do Solar Flares Affect Earth?

1. **Charged Particles:** Solar flares send out bursts of electrically charged particles called plasma. When these par-

ticles hit Earth's magnetic field, they create geomagnetic storms.

2. **High-Flying Trouble:** Satellites orbiting Earth can be damaged by solar flare energy, messing up cell phone and internet signals.

3. **Electrical Overload:** Power lines and transformers can experience sudden spikes, sometimes leading to blackouts in large areas.

Why Is This Important?

- **Space Weather Warnings:** By keeping an eye on solar flares, scientists and power companies can prepare, reducing damage to electronics and satellites.

- **Northern Lights Show:** Solar flare activity can make the aurora borealis (northern lights) glow even brighter and appear farther from the poles than usual.

- **Everyday Connections:** We depend on electricity for lights, computers, and more—so understanding the Sun's temper helps protect our modern world.

Fun Facts

- The Sun is over 90 million miles away, yet its flares can affect us in just a couple of days!
- Solar flares can be stronger than a billion atomic bombs going off at once, but luckily, Earth's magnetic field shields us from most of their harmful effects.
- The largest geomagnetic storm on record happened in 1859, called the **Carrington Event**. It made telegraph lines spark and even set papers on fire!

Try It Yourself

1. **Track the Sun:** Look up a "solar flare forecast" website (with an adult), just like a weather report. Note any upcoming flares and see if they coincide with northern lights sightings.
2. **Watch the Aurora:** If you live in or visit high-latitude regions, see if stronger flares cause brighter auroras. You can find live webcams online, too!
3. **Power Perspective:** Talk about what life would be like

if a huge solar flare knocked out electricity for a day. How would you cope without TV, phones, or lights?

Why It Matters

The Sun is our planet's life-giver, but it also packs a powerful punch. Solar flares remind us that Earth sits in a cosmic neighborhood where storms can fly through space and shake up our daily routines. By studying these flares, we learn how to protect our technologies—and appreciate the awe-inspiring power of the star that makes life possible!

25

Polar Bears in Your Cell Phone?

—*Arctic Metals and High-Tech Gadgets*

Have you ever thought about how your cell phone and polar bears might be linked? Many electronic devices need **rare metals** mined from places near the Arctic, where polar bears roam. From the coldest regions on Earth come precious minerals that end up in the smartphones, tablets, and gaming systems you use every day!

How Are Polar Bears and Phones Connected?

1. **Arctic Resources:** The Arctic region contains valuable metals like copper, nickel, and rare-earth elements, essential for circuit boards and batteries.

2. **Habitat Impact:** Mining for these metals can disrupt polar bear habitats, affecting the ice and waters they rely on for hunting seals.

3. **Global Demand:** As more people buy phones and gadgets, more resources are needed from around the world—including icy northern lands.

Why Is This Important?

- **Environmental Balance:** Mining activities must be carefully managed to avoid harming the Arctic's fragile ecosystem and the animals that live there.

- **Climate Change Link:** As global temperatures rise, Arctic ice melts sooner each year, making it even harder for polar bears to find food. This can be worsened if mining and shipping disturb their habitats further.

Fun Facts

- Polar bears have black skin under their white fur, which helps absorb heat from the Sun.

- Rare earth metals aren't actually rare—they're just hard

to extract. They're crucial for producing tech devices like smartphones and wind turbines.

- Some phone companies are exploring recyclable parts so fewer new metals are needed, which could reduce the need for more mines in sensitive regions.

Try It Yourself

1. **Recycling Round-Up:** Ask an adult about recycling old phones or laptops. This helps reuse metals instead of mining fresh resources.

2. **Polar Bear Research:** Read (with an adult) about polar bears' hunting habits and how melting ice forces them to swim longer distances for food.

Why It All Matters

Your cell phone is part of a global story connecting distant regions and wild creatures like polar bears. By learning about the Arctic's minerals and the animals that live there, we see how our technology choices affect ecosystems far away. It's a powerful reminder that our everyday devices link us not just to each other, but also to Earth's most remote and fragile environments.

26

Ancient Forests Beneath the Sea

—*Sunken Trees That Tell Old Stories*

Picture a lush green forest with towering trees, leafy branches, and winding roots. Now imagine that same forest buried under the ocean! In many coastal areas, scientists have discovered **submerged forests**—remnants of woodlands from thousands of years ago. Over time, rising sea levels and shifting shorelines covered these forests with water, preserving them like underwater time capsules.

How Do Forests End Up Underwater?

1. **Changing Coastlines:** Long ago, some shorelines were farther out to sea. As ice caps melted and sea levels rose, low-lying forests were flooded.

2. **Preserved by Sediment:** Sand, mud, and silt can settle on top of the trees, protecting wood from decay. This keeps fallen trunks and roots partly intact for centuries!

3. **Revealed by Tides:** In certain places, low tide reveals tree stumps jutting out of the sand, giving us a glimpse into an ancient ecosystem.

Why Is This Important?

- **Clues to the Past:** Studying these submerged forests helps scientists learn about ancient climates and how Earth's landscapes have shifted over thousands of years.

- **Fossil Discoveries:** Sometimes, archaeologists find prehistoric animal bones or human artifacts near these sunken woods, shedding light on how our ancestors lived.

- **Rising Seas Today:** Seeing how forests were submerged in the past helps us understand what might happen to coastal areas facing sea level rise now.

Fun Facts

- Some submerged forests are so old they date back to the last Ice Age, more than 10,000 years ago!
- In places like the Gulf of Mexico, scuba divers explore underwater cypress forests that look like spooky, moss-covered groves.
- Fossilized wood from these forests can still show growth rings, telling scientists about rainy and dry seasons long gone.

Try It Yourself

1. **Beach Walk:** If you visit a beach known for low tides, look for tree stumps or roots poking through the sand. They might be part of a submerged forest!
2. **Research the Past:** Check with an adult about local history. Were coastlines different near where you live? Some areas have changed dramatically over thousands of years.
3. **Ancient Diorama:** Create a model or diorama show-

ing a forest before sea levels rose, then cover parts of it with "water" made from blue cellophane or paper to see how it might have transformed.

Why It Matters

These underwater forests are like windows into Earth's history, showing us that our planet's surface is never truly fixed. Landscapes shift, seas rise, and entire forests can sink beneath the waves. By investigating these hidden woodlands, we learn how life adapts—and how our present choices about coastlines and climate may shape the forests of tomorrow.

27

Worms That Air-Condition the Soil

—Underground Tunnels Helping Roots Breathe

When you think of worms, you might imagine slimy creatures squiggling in the dirt. But did you know these humble helpers perform an amazing service for plants? Earthworms dig long, winding tunnels beneath the surface, letting air and water reach plant roots. It's almost like worms are little engineers, "air-conditioning" the soil so that crops and flowers can thrive!

How Do Worms Air-Condition the Soil?

1. **Tunnel Time:** As worms munch on decaying leaves and other organic matter, they create tunnels through the soil.

2. **Air and Water Flow:** These tunnels let fresh air and moisture move more easily underground, helping plants breathe and drink.

3. **Nutrient Mixing:** Worms also leave behind nutrient-rich waste called castings, which fertilize the soil and feed hungry plants.

Why Is This Important?

- **Root Boost:** When roots can spread out in well-aerated soil, plants grow stronger and produce more flowers or fruit.

- **Healthy Gardens:** Worms help break down dead leaves and other scraps, turning them into natural fertilizer for your garden or local park.

- **Eco-Engineers:** Soils full of active worms are less prone to erosion and more resistant to flooding, protecting landscapes from damage.

Fun Facts

- Earthworms don't have lungs; they breathe through their skin, which must stay moist so they can absorb oxygen.
- Some worm species can grow over 3 feet long in certain parts of the world—imagine finding one of those in your garden!
- If a worm's tail is cut off but its vital organs remain in the front segment, it may survive and regenerate part of its tail.

Try It Yourself

1. **Worm Observation:** Create a mini worm habitat (with an adult) by layering soil and sand in a clear container. Add leaves on top for food. Watch how worms mix the layers.
2. **Garden Check:** Gently dig a small hole in your yard or

a planter. See if you spot any worms—or their tunnels.

3. **Compost Crew:** If you have a compost bin, worms will likely show up naturally. They're great at turning kitchen scraps into rich soil!

Why It Matters

Earthworms may seem simple, but they're vital for healthy plants, gardens, and whole ecosystems. By creating tunnels that let air and water flow, they make the ground a better home for roots and countless other organisms. These underground "air conditioners" show that some of nature's most important jobs are done by the smallest, squirmiest workers of all!

28

Peacock Feathers and Rainbows

—The Science of Shimmering Colors

When a peacock spreads its gorgeous tail, the feathers gleam with dazzling blues, greens, and purples that shift as the bird moves. But have you ever wondered why peacock feathers seem to glow like rainbows? It's not just simple color—there's a special trick at work called **structural coloration**, which bends and reflects light in magical ways!

How Do Peacock Feathers Create Such Brilliant Hues?

1. **Tiny Structures:** Each feather has microscopic ridges or layers that split and reflect different wavelengths of

light.

2. **Color Changes:** As the peacock moves, the angle of the light changes. This makes the colors look like they're shifting, almost like a rainbow.

3. **Mating Signals:** Male peacocks show off their shimmering feathers to attract females, who often prefer the brightest, flashiest displays.

Why Is This Important?

- **Nature's Artwork:** Peacock feathers show how clever nature can be, creating colors by bending light instead of just using pigments like paint.

- **Copying Nature:** Scientists study structural coloration to develop better screens, paints, and even materials that can change color when moved.

- **Biodiversity:** Peacocks and their feathers remind us of the incredible variety of ways animals adapt to attract mates or ward off predators.

Fun Facts

- Some butterflies and beetles also use structural coloration, giving them an iridescent look similar to a peacock's feathers.
- A peacock's tail can contain more than 200 shimmering feathers!
- In many parts of the world, peacocks symbolize beauty, pride, and even immortality in traditional stories.

Try It Yourself

1. **Feather Inspection:** If you find a feather (or buy one at a craft store), hold it in the light and move it around. Notice how the color shifts at different angles.
2. **Rainbow Science:** Shine a flashlight through a clear CD (with an adult's help). See how the grooves reflect light in rainbow patterns—kind of like a peacock's feather.
3. **Art Project:** Use crayons or markers to draw a peacock feather. Shade it with blues, greens, and purples, then

add shiny details to mimic iridescence.

Why It Matters

Peacock feathers don't just look cool—they show us how intricate and creative nature can be when it comes to color. By bending light instead of using pigment, these shimmering displays captivate us and spark new ideas in science and art. It's yet another reminder of how everyday wonders—like a peacock's plumage—can inspire us to see the world in brighter, more brilliant ways!

29

Mosquitoes and the Food Chain

—Pesky Pests with a Purpose

You've probably felt that sting on a summer evening—a **mosquito** bite! While these buzzing insects can be annoying (and even carry diseases), mosquitoes also play an unexpected role in nature. As food for fish, birds, and other creatures, they form a **key link** in the food chain. It's a reminder that even the most irritating animals can have an important job in the environment.

How Do Mosquitoes Fit Into the Food Chain?

1. **Water Babies:** Mosquitoes lay eggs in still water—puddles, ponds, or even old tires. The larvae that hatch become tasty treats for fish and amphibians.

2. **Buzzing Bites:** Adult mosquitoes feed on nectar for energy. Only female mosquitoes bite mammals for blood to help them lay eggs—annoying for us, but helpful for their species' survival.

3. **Bird Feast:** Many birds, bats, and dragonflies eat adult mosquitoes, reducing their numbers and turning the pests into a protein-packed meal.

Why Is This Important?

- **Balancing Act:** If mosquitoes disappeared completely, some animals would lose a big part of their diet, possibly throwing ecosystems off-balance.

- **Nutrient Cycling:** Mosquito larvae help break down organic matter in the water, recycling nutrients that feed other aquatic life.

- **Scientific Clues:** Scientists study mosquito behavior and genetics to learn about disease spread, guiding public health efforts to keep us safe.

Fun Facts

- Mosquitoes have been around for over 100 million years—long before humans existed!
- There are more than 3,000 mosquito species worldwide, but only a few bother people regularly.
- Male mosquitoes don't bite; they stick to sipping flower nectar.

Try It Yourself

1. **Mosquito Watch:** Check for standing water in your yard—like buckets or old plant trays—and empty them to reduce mosquito breeding spots.
2. **Nature Observation:** Visit a pond or marsh (with an adult) and see if you can spot tiny mosquito larvae wiggling in the water.
3. **Food Chain Diagram:** Draw a simple food chain, placing mosquito larvae at the center. Show fish eating larvae, then herons or bats eating the fish or adult mosquitoes.

Why It Matters

Mosquitoes might make you itch, but they're a crucial food source for many animals. By annoying us and feeding others, they link various parts of the ecosystem together. Understanding their role helps us keep nature's balance in mind—even when we're swatting away at these tiny, buzzing pests.

30

Beach Glass from Trash to Treasure

—Ocean Waves Turn Bottles into Jewels

Ever picked up a smooth, colorful piece of glass on the beach and wondered how it got there? That shiny pebble might have started as a **broken bottle** or **shard of window glass** thrown into the ocean. Over time, waves and sand grind down the sharp edges, turning garbage into gorgeous, frosty bits of "sea glass." It's a magical transformation that reminds us of nature's power to reshape our discarded items into something beautiful.

How Does Glass Become Sea Glass?

1. **Shattered Trash:** Glass objects like bottles, jars, or

windows sometimes end up in the ocean, either by accident or careless littering.

2. **Wave Action:** Constant tumbling in the surf wears away the glass's sharp edges, smoothing it into rounded shapes.

3. **Frosty Finish:** Saltwater and sand also etch a cloudy, matte surface onto each piece, giving sea glass its distinct glow.

Why Is This Important?

- **Ocean Cleanup:** Although it starts as litter, sea glass is evidence of how the ocean can naturally recycle, given enough time and movement.

- **Unique Souvenirs:** Each piece of sea glass has a one-of-a-kind color and shape, making it a small treasure hunt for collectors.

- **Environmental Message:** Even though nature can transform trash, it's still better to **reduce, reuse, and recycle** before items ever reach our oceans.

Fun Facts

- Sea glass comes in many colors, but brown, green, and clear are most common because they come from common bottles. Rare colors—like red, turquoise, or pink—can be highly prized by collectors!
- Some beach glass dates back many decades or even a century, especially if it came from old shipwrecks or abandoned dump sites.
- In some places, beaches known for sea glass are protected, encouraging visitors to enjoy looking but leave most pieces behind.

Try It Yourself

1. **Treasure Hunt:** If you visit a beach known for sea glass, look along the high tide line. Keep your eyes peeled for colorful shapes among the sand and pebbles.
2. **Crafty Creations:** Make sea glass art (with an adult's help). Glue pieces to a picture frame or try stringing them into a necklace.

Why It Matters

Beach glass shows how nature's forces—waves, sand, and salt—can gently transform human trash into something lovely. Each smooth fragment is a reminder that our planet can recycle and renew itself, but only to a point. By being careful with our waste, we help keep the ocean healthier and make it easier for nature's own cleanup crew to do its job!

31

The Busy Bee and Your Blue Jeans

—*Cotton Plants Need Fuzzy Pollinators*

Next time you slip on a pair of jeans, think about the **bees** who helped make them possible! Cotton plants—the source of the fibers in denim—depend on pollinators like bees to help their flowers turn into cotton bolls. Without these hard-working insects, our closets might look very different. It's a fun reminder that the clothes we wear often start with a buzzing visit in the fields.

How Do Bees Help Cotton Plants Grow?

1. **Flower Pollen:** Bees land on cotton flowers to sip nectar. While they're busy drinking, pollen sticks to their

fuzzy bodies.

2. **Flower to Flower:** As a bee flies to another cotton blossom, some of that pollen rubs off, fertilizing the flower so it can grow a cotton boll.

3. **Soft Fibers:** Once the flower is pollinated, it forms a pod filled with fluffy cotton fibers. Over time, these fibers are harvested, spun into thread, and woven into the denim fabric we know and love.

Why Is This Important?

- **Clothing Connection:** No bees could mean fewer cotton bolls, which could lead to shortages or higher prices for cotton-based clothing.

- **Ecosystem Support:** Bees that pollinate cotton also help many other plants in the area, creating a healthier environment for everyone, including us!

- **Farming Partnerships:** Farmers often plant wildflowers near cotton fields to attract more bees, improving pollination and boosting crop growth.

Fun Facts

- Honeybees can visit thousands of flowers in a single day—imagine the miles they cover!
- Cotton plants produce pretty white or yellowish flowers that change color as they mature.

Try It Yourself

1. **Plant a Pollinator Garden:** Ask an adult to help you grow nectar-rich flowers. Bees visiting your garden might also help any local farms growing cotton or other crops.
2. **Check Your Tags:** Look at the labels on your jeans and shirts to see if they're made from 100% cotton or a blend. Either way, bees likely played a role.

Why It Matters

The buzzing of bees isn't just background noise—these tiny insects make a huge impact on what we wear and eat. Every time you throw on your favorite jeans, you're wearing a piece of nature's teamwork between bees and plants.

32

Desert Plants That Make Clouds

—Hot Air Rising to Make Rain Far Away

When you think of deserts, you might picture endless sand and scorching sun—but certain plants in these dry lands do more than just survive. They also release water vapor that rises into the sky, sometimes helping to **form clouds** far beyond the desert's borders. It's like the desert is secretly sharing its moisture with other parts of the world!

How Do Desert Plants Make Clouds?

1. **Thirsty Leaves:** Desert plants like cacti and succulents store water in thick leaves or stems. They carefully release small amounts of water vapor during cooler times

of the day.

2. **Evapotranspiration:** This process combines evaporation (water turning into vapor) and transpiration (plants breathing out water), sending moisture into the air.

3. **Wind Carrying Vapor:** Strong desert winds can blow this vapor upward. Eventually, it can cool down, condense, and form clouds that drift to other regions, bringing rain where it's needed.

Why Is This Important?

- **Water Sharing:** Even though deserts are dry, they can send moisture to places that might be far away, connecting ecosystems.
- **Climate Influence:** Clouds reflect some of the Sun's heat back into space. By creating more clouds, desert plants can subtly affect Earth's temperature.
- **Survival Trick:** Plants in deserts show incredible adaptations, proving that life can find clever ways to manage

scarce resources.

Fun Facts

- Some cacti open their pores at night (when it's cooler) to release water vapor without losing too much moisture.
- The giant saguaro cactus can store thousands of gallons of water and slowly release it over time.
- Desert winds, like the **Harmattan** in Africa or the **Santa Ana** in California, can carry dust and moisture across long distances.

Try It Yourself

1. **Plant Detective:** Look at pictures of different desert plants. Notice how many have thick, waxy surfaces or spines instead of leaves—this helps reduce water loss.
2. **Local Weather Patterns:** Check how far clouds typically travel in your region. Imagine desert moisture being part of that journey!

3. **Mini Greenhouse:** Place a small succulent in a clear container with some water. Watch how droplets form on the sides, showing how plants and water vapor interact.

Why It Matters

Desert plants remind us that even the driest places on Earth have ways of contributing to global weather patterns. By releasing water vapor and helping form clouds, these hardy survivors connect distant climates in ways we might never suspect. It's a shining example of how every ecosystem, no matter how harsh, plays a part in our planet's grand water cycle.

33

PENGUINS AND AIRPLANES

—Ice Clues That Guide Plane Routes

Penguins waddle across icy landscapes at the bottom of the world, far from the busy airports most of us use. So, how do these flightless birds connect to jet planes flying in the skies above? Surprisingly, penguin colonies help scientists track **sea ice changes**, which can influence the weather conditions pilots consider when mapping safe flight routes—especially those passing near polar regions!

HOW DO PENGUINS HELP AIR TRAVEL?

1. **Colony Locations:** By monitoring where penguins nest and how their colonies move, researchers learn about shifting sea ice. If the ice shrinks, penguins may

relocate or struggle to find food.

2. **Weather Insights:** Changes in sea ice can affect ocean temperatures and global wind patterns, potentially altering flight routes over or near the poles.

3. **Polar Pathways:** Airlines sometimes take "great circle" routes (the shortest distance around the globe), which can go near polar areas. Knowing ice and weather conditions is crucial for safety and fuel planning.

Why Is This Important?

- **Global Weather Links:** When sea ice melts or expands, it can change ocean currents and jet streams—affecting flights far beyond the poles.

- **Wildlife Health:** Healthy penguin populations indicate stable ice conditions, which means fewer surprises for long-distance flight routes.

- **Team Science:** Studying penguins combines biology, climate science, and aviation research, showing how different fields work together to keep skies safe.

Fun Facts

- Some airlines occasionally fly routes over the Arctic or near Antarctica to shorten travel times, depending on the season and weather.
- Emperor penguins can huddle together to withstand brutally cold Antarctic winters—temperatures that can drop below -60°F!
- Penguin feathers are so well-insulated that these birds can take a dip in freezing water and still keep warm.

Try It Yourself

1. **Penguin Monitoring:** Look up satellite images (with an adult) showing penguin colonies. Notice how data about their locations is used in climate studies.
2. **Flight Map Exploration:** Check out a globe or online flight tracker to see how some planes go over polar regions—these are often the fastest routes between continents.
3. **Climate Connection:** Research (with an adult) how

melting sea ice can change ocean currents. Discover how that might affect your local weather—even if you live far from the poles!

Why It Matters

You might not think "penguins" when you board an airplane, but these cute, chilly birds offer vital clues about our planet's changing climate—clues that can help pilots plan flights. It's another reminder that **no place is too remote** to impact our daily lives and that wildlife, weather, and human technology all intertwine in surprising ways!

34

Coral That Grows from a Volcano

—*Lava Lands Turn into Ocean Reefs*

Close your eyes and picture a **volcano** spewing fiery lava into the ocean. That hot, molten rock might seem like the last place for anything to live, but guess what often happens over time? The cooled lava can become a home for **tiny coral polyps** to settle and grow. Before long, a brand-new coral reef can spring up where a volcano once raged, showing how nature transforms destruction into vibrant, living habitats!

How Do Volcanoes Help Corals Grow?

1. **Cooling to Rock:** When lava hits ocean water, it cools and hardens into rocky surfaces—perfect places for baby corals to attach and begin building reefs.

2. **New Ocean "Real Estate":** These fresh rocky areas don't have many competing organisms, so corals find space to spread out and form whole reefs.

3. **Ecosystem Building:** As corals grow, they attract fish, sea turtles, and other marine life, creating a bustling underwater community around the old volcano site.

Why Is This Important?

- **Natural Renewal:** It's a reminder that volcanic eruptions, while destructive, can lead to new environments for sea life.

- **Coral Biodiversity:** More reefs mean more habitats for colorful fish and other marine creatures, helping preserve ocean biodiversity.

- **Coastal Protection:** Coral reefs act like underwater barriers, helping protect coastlines from strong waves and storms.

Fun Facts

- Some volcanic islands, like those in Hawaii, have black

sand beaches made from tiny lava fragments that gradually weather into sand.

- Corals are actually tiny animals, not plants! Each coral polyp works together with algae called zooxanthellae, which help them get energy from sunlight.
- Over time, coral reefs can grow large enough to form atolls—ring-shaped islands surrounding a lagoon where a volcano once stood.

Try It Yourself

1. **Volcanic Island Research:** With an adult's help, look up volcanic island chains like the Galápagos or Hawaii. Check how reefs have formed around them.
2. **Colorful Corals:** Draw or paint a scene showing new coral polyps attaching to cooled lava rock. Add vibrant fish and imagine the reef's growth over time!

Why It Matters

Volcanoes may seem scary, but they can set the stage for remarkable life under the sea. By creating fresh surfaces for corals to colonize, eruptions ultimately foster thriving marine ecosystems. It's a breathtaking example of how our planet renews itself, turning fiery destruction into a cradle for new life.

35

THE "FARTING" SWAMP GAS

—Bubbling Marshes That Affect Climate

Venture into a marshy swamp, and you might see bubbles rising to the water's surface—some people jokingly call them "swamp farts!" These bubbles are actually **methane**, a gas produced by bacteria living in the mud below. While it may sound silly, this "farting" swamp gas can have a **big impact** on our planet's climate, showing us again how even smelly gases connect us all.

HOW DOES SWAMP GAS FORM?

1. **Underwater Breakdown:** In swampy, waterlogged soil, oxygen is limited. Certain bacteria break down

dead plants and animals, releasing methane gas.

2. **Bubble Up:** Methane collects in pockets and eventually escapes through the water's surface, often appearing as bubbles.

3. **Atmospheric Journey:** Once in the air, methane can trap heat (like carbon dioxide), influencing global temperatures.

Why Is This Important?

- **Climate Balance:** Methane is a powerful greenhouse gas, meaning too much of it can heat our atmosphere. Swamps and marshes contribute a natural source of methane.

- **Wetlands Health:** Healthy wetlands do more than release methane—they also filter water, prevent floods, and support countless species.

- **Feedback Loops:** As global temperatures rise, swampy areas may expand, releasing more methane. Scientists study these loops to predict future climate changes.

Fun Facts

- Methane is odorless on its own; the rotten-egg smell often comes from other sulfur compounds in marshy areas.
- In very cold regions, trapped methane can ignite and burn if released suddenly, sometimes causing eerie flickering lights in bogs—legend calls them "will-o'-the-wisps."
- Wetlands, including swamps and marshes, are among the most productive ecosystems on Earth, teeming with fish, birds, insects, and more.

Try It Yourself

1. **Nature Walk:** Visit a local marsh or swamp (with an adult). Look for bubbles on the water's surface, especially where the mud is thick.
2. **Wetland Research:** Explore (with an adult) online maps of wetland areas near you. Are they expanding or shrinking over time? Why might that matter?

3. **Draw a Food Web:** Include plants, bacteria, insects, and bigger animals like alligators or birds. Show how their interactions eventually lead to methane bubbles.

Why It Matters

Those funny little bubbles show that Earth's systems are deeply intertwined. Swamps might look muddy and messy, but they play a vital role in balancing our climate—and providing homes for countless creatures. It's one more reminder that even "farting" marshes fit into the grand puzzle of life on our planet.

36

Elephants Saving Butterflies

—Big Feet That Clear Paths for Little Wings

In the vast grasslands and forests of Africa and Asia, you'll often spot mighty elephants lumbering through. With their huge bodies and powerful trunks, these gentle giants knock down trees, trample thick undergrowth, and even dig up waterholes. Surprisingly, all this "destruction" can actually help tiny creatures like **butterflies** thrive. By clearing openings and stirring up nutrients, elephants pave the way for new plants and flowers, feeding flutters of beautiful pollinators!

How Do Elephants Help Butterflies?

1. **Clearing Paths:** As elephants move, their massive bod-

ies push over shrubs and small trees, creating sunny patches where wildflowers can sprout—perfect butterfly food.

2. **Seed Spreaders:** Elephants eat fruits and seeds, then drop them far away in their dung (poop). This fertilized seedbed helps new plants, providing more nectar for butterflies.

3. **Water Holes:** Using trunks and feet, elephants dig for water during dry seasons. When these spots fill with rain, butterflies and other creatures have fresh drinking sources.

Why Is This Important?

- **Ecosystem Architects:** Elephants are known as "ecosystem engineers," shaping their habitats in ways that benefit many other species.

- **Butterfly Bonanza:** More flowers mean more nectar, supporting butterflies, bees, and other pollinators that help plants reproduce.

- **Food Chain Support:** Healthy butterfly populations feed birds, reptiles, and even some mammals, keeping nature's web strong.

Fun Facts

- Elephants can eat hundreds of pounds of vegetation each day, making them walking lawnmowers in the wild.
- A single elephant can poop out seeds that grow into hundreds of new plants over its lifetime!
- Butterflies taste with their feet—so having fresh flowers and puddles to land on means a tasty buffet.

Try It Yourself

1. **Habitat Drawing:** Sketch a savanna or forest with elephants knocking down bushes and butterflies visiting new wildflowers. Notice how each action leads to another's benefit.
2. **Pollinator Garden:** If you can't bring an elephant home, you can still help butterflies by planting native

wildflowers that they love.

Why It Matters

Elephants may seem too big to connect with something as delicate as butterflies, yet their actions create sunny meadows and scattered seeds for these small fliers. It's a powerful reminder that every creature, no matter its size, plays a vital role in shaping life on Earth. From the mightiest elephants to the tiniest butterflies, we're all in this together!

37

The Racing Seeds and Helpful Rivers

—Streams Carry Future Forests

Picture a tiny seed dropping from a tree and splashing into a river. Instead of sinking or vanishing, it floats along, carried by the current. This seed might travel miles downstream before finally settling in new soil and sprouting into a young sapling. Through this natural "shipping system," rivers help **spread** seeds and create fresh patches of forest far from where they began!

How Do Rivers Help Seeds Travel?

1. **Falling In:** Seeds from trees and plants near a riverbank may drop directly into the water—sometimes with a little help from wind or curious animals.

2. **Floating Journey:** Certain seeds have air pockets or waterproof coats that help them float for days or weeks, giving them plenty of time to find new shores.

3. **Perfect Landing:** When the current slows or the river bends, seeds may wash ashore. If the soil is right, they can germinate and grow into a new tree or plant.

Why Is This Important?

- **Forest Expansion:** By spreading seeds far downstream, rivers help forests expand naturally, bringing shade and nutrients to new areas.

- **Biodiversity Boost:** Each new patch of trees creates a habitat for birds, insects, and other wildlife, increasing the variety of life along the river.

- **Erosion Control:** When more trees sprout, they anchor the soil with their roots, reducing erosion along riverbanks and keeping waterways healthier.

Fun Facts

- Some seeds, like coconut seeds, can float across entire

oceans, helping coconut palms grow on distant islands.

- Rivers are a key part of the **seed dispersal** process, joining forces with animals, wind, and even people in spreading new plant life.
- Certain fish might nibble on seeds, carrying them even further. If the seeds survive the trip through a fish's body, they can sprout later!

Try It Yourself

1. **Seed Float Experiment:** Fill a bucket with water and drop in different seeds (sunflower, bean, etc.). Which ones sink or float? Imagine how far each might travel in a river.

2. **River Map:** Look at a local river map (with an adult). Identify spots where the current might slow, creating a good place for seeds to settle.

3. **Draw a Seed's Journey:** Illustrate a seed's path from a tall tree, into a river, then onto a new stretch of land where it becomes a sapling.

Why It Matters

Rivers do more than provide water—they also act like nature's mail carriers for seeds. By ferrying new plants to distant shores, they shape ever-changing landscapes and support a growing web of life. It's one more sign of how Earth's systems—forests, rivers, and all living creatures—stay linked in a continuous cycle of renewal.

38

Your Breath and the Rainforest

—*Trading Oxygen for Carbon*

Did you know that the air you breathe is connected to the **rainforest** in places you may never have visited? Plants and trees—especially those in lush tropical forests—use sunlight to turn carbon dioxide (CO_2) into oxygen through **photosynthesis**. Meanwhile, humans (and animals) breathe in oxygen and exhale carbon dioxide. It's like an invisible handshake across the planet, where **your breath** and rainforest trees **trade** essential gases to keep life thriving!

How Does This Exchange Work?

1. **Trees Inhale CO_2:** During photosynthesis, leaves absorb carbon dioxide from the air, using it to create sug-

ars for energy and growth.

2. **Trees Exhale Oxygen:** As a byproduct of photosynthesis, trees release oxygen. This becomes part of the air we breathe.

3. **You Breathe Out CO_2:** When you exhale, you send carbon dioxide back into the atmosphere—some of which makes its way to rainforests and other green spaces.

Why Is This Important?

- **Global Air Filter:** Rainforests act like giant lungs, helping to maintain the balance of oxygen and carbon dioxide on Earth.

- **Climate Control:** By storing carbon, rainforests help cool the planet. More trees generally mean less CO_2 in the air, slowing climate change.

- **Habitats and Diversity:** Rainforests are home to more than half of all plant and animal species on land, making them vital for biodiversity.

Fun Facts

- The Amazon Rainforest alone produces around 20% of the world's breathable oxygen—though it's also important to remember oceans produce even more!
- One large tree can supply enough oxygen for four people each day.
- Trees also help filter out pollutants, keeping our air fresher and healthier.

Try It Yourself

1. **Breathing Exercise:** Take a deep breath outside near a tree. Imagine the invisible journey that oxygen took from a leaf, through your lungs, and back out as CO_2.
2. **Plant a Tree:** Talk to your family or school about planting a native tree. Each new sapling is a mini oxygen factory in your neighborhood!
3. **Rainforest Research:** Look up (with an adult) the biggest rainforests in the world, like the Amazon or the Congo Basin, and learn about how they're being

protected—or threatened.

Why It Matters

Your simple act of **breathing** links you to green life across the globe, from the tree in your backyard to the towering giants in distant rainforests. Each inhale and exhale is part of a grand exchange that keeps our atmosphere balanced and supports countless living things. It's another beautiful reminder that our planet truly breathes as one.

39

Baby Spiders That Fly on Silk

—*Tiny Web Trampolines Riding the Wind*

Imagine a baby spider climbing to the top of a branch, lifting its tiny abdomen, and releasing a silken thread into the breeze. All of a sudden—**whoosh!**—the wind catches the silk and whisks the spider away. This high-flying trick, called **ballooning**, helps baby spiders travel to new places, spreading spider populations far and wide.

How Do Spiders Balloon?

1. **Silk Spinners:** Baby spiders, called spiderlings, release thin strands of silk from their spinnerets.

2. **Catching the Wind:** They wait for a light breeze to tug on the silk, then let go of their perch, floating off into the air.

3. **Long-Distance Flight:** With the wind's help, some spiderlings can drift for miles—even across oceans—before landing in a fresh spot to start spinning webs of their own.

Why Is This Important?

- **New Homes:** Ballooning spiders can escape crowded areas or find better places with more insects to eat.

- **Web of Life:** Spiders control insect populations, helping farms and gardens by eating pests. Spreading out ensures they don't all compete in the same spot.

- **Nature's Adaptability:** Ballooning shows how even tiny creatures use clever strategies to survive and explore new environments.

Fun Facts

- Scientists have found spiderlings on weather balloons

and even at high altitudes, proving these little travelers can hitch rides in the sky!

- Some spiders can sense electrical charges in the air, using them to help lift off when conditions are perfect.
- Not all spider species balloon, but many do—especially in their earliest stage of life.

Try It Yourself

1. **Sky Watch:** On a calm, sunny day, look closely among garden plants or fence posts. You might spot tiny threads of silk floating around!
2. **Wind Experiment:** Lightly toss small pieces of thread or tissue into a gentle breeze. Notice how they ride the air—similar to how spiderlings travel.
3. **Spider Field Guide:** With an adult, look up local spider species. See which ones are known to balloon and how far they can drift.

Why It Matters

Baby spiders might seem too small to conquer the world, but their silk ballooning shows that size isn't everything. By catching the wind, they can travel to distant places, spread out, and keep insect populations in check. It's yet another example of how creative and resourceful nature can be—lifting tiny spiders skyward and stitching our planet's ecosystems together.

40

The Bird That Plants Fruit Trees

—Seeds Inside a Traveling Tummy

Have you ever thought about how fruit trees end up scattered across a forest? One big reason is that **fruit-eating birds**, like toucans or parrots, swallow seeds and then fly away. After a while, those seeds pass through the bird's digestive system and drop to the ground—often **far** from the original tree. In this way, birds become **gardeners of the wild**, planting new fruit trees wherever they go!

How Do Birds Help Plant Trees?

1. **Fruit Feasts:** Birds snack on berries, nuts, and fruits. Inside these treats are seeds waiting to be spread.

2. **Fly Away:** The birds travel from one place to another, carrying seeds in their bellies.

3. **Natural Drop-Off:** Eventually, the seeds exit the bird and land on fresh soil. With a little luck (and rain), those seeds can sprout into new saplings.

Why Is This Important?

- **Forest Diversity:** Birds can carry seeds across long distances, helping forests spread and stay genetically varied.

- **Pollination Partners:** While not strictly pollinators, fruit-eating birds also help some plants by scattering their seeds where conditions may be ideal for growth.

- **Animal Food Chain:** Fruit trees grown from these seeds can one day feed more birds, monkeys, and even people. It's a cycle of giving!

Fun Facts

- Some tropical birds have specialized beaks to open tough nuts, cracking them just enough for the seed to be ingested or dropped.

- Large birds like hornbills can carry big seeds; smaller birds help spread tiny ones. Each bird size plays a role in the forest.

- Seeds often survive the bird's digestive juices, sometimes even sprouting better after being softened inside a bird's stomach!

Try It Yourself

1. **Bird Watch:** With an adult, observe local fruit-eating birds—like robins, cardinals, or grosbeaks. Notice if they carry fruit away from the tree where they found it.

2. **Seed Sprouts:** If you eat fruit with seeds (like apples or berries), see if you can plant a few in a small pot. Imagine a bird doing this in the wild!

3. **Food Web Drawing:** Make a mini-ecosystem sketch showing a fruit tree, a bird eating its fruits, and a new seedling sprouting far away.

Why It Matters

Birds don't just add color and music to forests; they act as nature's **seed couriers**, making sure new generations of trees take root in distant spots. It's one more way that creatures big and small work together to keep our planet lush, green, and ever-changing—proving that **even a simple fruit snack** can grow into a mighty tree in the grand dance of life.

41

Jellyfish and Robot Science

—*Ocean Blobs Inspiring High-Tech Solutions*

You might look at a **jellyfish** drifting in the ocean and think it's just a wiggly blob of jelly. But did you know these graceful creatures have inspired inventors to create **new kinds of robots**? That's right—jellyfish, with their smooth pulsing motions, provide clues for designing underwater robots that can move gently through water without harming fish or coral reefs!

How Do Jellyfish Help Invent Robots?

1. **Smooth Swimming:** Jellyfish swim by gently pushing water behind them. Engineers study this movement to build robots that can glide without using sharp pro-

pellers.

2. **Flexible Body:** A jellyfish's bell (the round top part) changes shape with each pulse. Robotic designs that mimic this flexibility can explore tight spaces underwater.

3. **Low Energy:** Jellyfish don't waste much energy to move. Robots that copy their efficiency could use less power and run for longer periods under the sea.

Why Is This Important?

- **Ocean Exploration:** Soft-bodied robots can explore fragile environments like coral reefs without causing damage.

- **Wildlife Protection:** Quiet, gentle robots reduce noise pollution, making them friendlier to marine life.

- **New Technologies:** Research on jellyfish movement could lead to medical devices, rescue robots, or improved water filters—beyond just ocean science!

Fun Facts

- Some jellyfish are **bioluminescent**, meaning they glow in the dark. Researchers also study this for cool lighting and medical imaging ideas.
- Jellyfish have been around for **hundreds of millions of years**, outliving dinosaurs!
- A group of jellyfish can be called a "smack," "bloom," or "swarm," and they sometimes gather in huge numbers to feed or drift with currents.

Try It Yourself

1. **Swimming Mimic:** Watch slow-motion videos of jellyfish swimming (with an adult). Notice how they pulse and contract, then imagine how a robot might copy those moves.
2. **Design a Robo-Jelly:** Draw a simple plan for a robot that uses a soft, bell-like shape. Label parts for sensors, batteries, and a flexible "body."
3. **Science News:** Look up (with an adult) recent news

about soft robotics or jellyfish-inspired inventions. You might be surprised at how many ideas come from these ocean drifters!

Why It Matters

Jellyfish may look simple, but their gentle, efficient swimming style is a treasure chest of ideas for human technology. By observing these silent ocean dwellers, scientists and engineers learn to build kinder, gentler robots that protect our seas. It's another reminder that **nature's designs**, refined over millions of years, can lead us to smarter, more sustainable innovations today.

42

Desert Winds Feed the Amazon

— *Dust Crossing the Atlantic Ocean*

It might sound impossible, but **dust** from the Sahara Desert in Africa travels all the way across the **Atlantic Ocean** to help **fertilize** the lush **Amazon Rainforest** in South America! This cross-continental connection shows how the driest places on Earth can give a helping hand to one of the wettest and most vibrant ecosystems out there.

How Does Sahara Dust Reach the Amazon?

1. **Powerful Winds:** Strong desert winds kick up tiny dust particles from the Sahara's sandy surface.

2. **Long Flight:** High-altitude air currents carry this dust

in huge clouds across the ocean, sometimes traveling thousands of miles.

3. **Rainforest Delivery:** When the dust settles, it drops onto the Amazon, where it provides nutrients—like phosphorus—that help rainforest plants grow.

Why Is This Important?

- **Global Fertilizer:** Without these nutrient-rich dust particles, the Amazon soil might be less fertile, affecting countless plant and animal species.
- **Climate Influence:** Dust particles can reflect sunlight and affect cloud formation, subtly changing weather patterns on both sides of the ocean.
- **Planetary Ties:** This dust highway shows that even places separated by vast oceans are connected by wind, weather, and Earth's constant cycling of materials.

Fun Facts

- The Sahara Desert is the **world's largest hot desert**, covering much of North Africa.

- Scientists estimate that over **20 million tons** of Saharan dust may land in the Amazon each year!
- The dust contains minerals left behind from an ancient lakebed in Chad called Lake Megachad, which dried up long ago.

Try It Yourself

1. **World Map Discovery:** Look at a globe or world map (with an adult) and trace a line from the Sahara to the Amazon. Notice how far the dust must travel!
2. **Dust in the Air:** On a dry, windy day, look at sunbeams coming through a window. You might see dust swirling around—even on a much smaller scale, that's how it travels.
3. **Science News Check:** Sometimes, Saharan dust clouds reach places like the Caribbean or the southern United States. See if you can find reports of dusty skies in those regions!

Why It Matters

The story of desert dust feeding a rainforest proves that **no corner of Earth stands alone**. From Africa's sandy dunes to the Amazon's green canopies, invisible wind currents link ecosystems separated by vast waters. It's another astonishing reminder of how our planet's living systems depend on each other—across oceans, continents, and deserts alike.

43

Otters, Kelp, and Carbon Clouds

—Ocean Furry Friends That Fight Climate Change

Imagine a forest—but underwater! **Kelp forests** can grow taller than houses, providing shelter for fish, crabs, and countless ocean creatures. And guess who helps keep these watery woodlands healthy? **Sea otters!** By eating sea urchins that gobble kelp, these playful, furry mammals allow kelp forests to flourish. That's not all: thriving kelp forests can remove **carbon dioxide (CO_2)** from the water, helping combat climate change.

How Do Otters Help Kelp Grow?

1. **Sea Urchin Snack:** Sea urchins love munching on kelp roots. Too many urchins can wipe out entire kelp

forests, creating "urchin barrens."

2. **Otter Appetite:** Sea otters feast on urchins, keeping their population under control. With fewer urchins, kelp can regrow and spread.

3. **Carbon Capture:** Healthy kelp forests soak up CO_2—just like land forests—reducing the amount of this greenhouse gas in our oceans and atmosphere.

Why Is This Important?

- **Biodiversity:** Kelp forests are like underwater jungles, hosting a wide variety of marine life. Keeping them healthy means more homes for fish, seals, and other creatures.

- **Climate Control:** By storing carbon in their tall, leafy fronds, kelp forests help slow global warming, working alongside land forests to clean our air.

- **Ecosystem Balance:** Otters are considered a **keystone species**—without them, the entire kelp forest might collapse.

Fun Facts

- Sea otters sometimes use **rocks** as tools to crack open shellfish—one of the few animals known to use tools!
- Kelp can grow up to **two feet a day**, reaching lengths of over 100 feet in some areas.
- Otters have the densest fur of any mammal, helping them stay warm in chilly waters without blubber.

Try It Yourself

1. **Otter Observation:** With an adult, search for videos of sea otters cracking open urchins or clams. Notice their clever paws and how relaxed they look floating on their backs!
2. **Kelp Exploration:** If you live near the coast, visit a tide pool or kelp forest area. Look for snails, crabs, and fish that call kelp home.
3. **Draw a Food Web:** Sketch a simple diagram showing how kelp, sea urchins, and otters connect. Label where CO_2 might enter or leave the system.

Why It Matters

Sea otters may seem like just adorable fluff-balls, but they play a **huge** part in maintaining kelp forests—a vital force against climate change. When otters thrive, kelp can spread, capturing more carbon and supporting rich marine ecosystems. It's an inspiring story of how one small predator can tip the balance back toward a healthy, vibrant ocean world.

44

The Surprising Power of Earthworms

—*Slime Trails That Turn Leaves into Rich Soil*

If you've ever dug in a garden and found a wriggling earthworm, you might have wondered—what good do these slimy creatures really do? As it turns out, **earthworms** are some of the **hardest-working recyclers** on the planet! By munching dead leaves and other organic material, then leaving behind nutrient-rich castings (worm poop), they help turn debris into **fertile soil**. It's like a never-ending compost factory beneath your feet!

How Do Earthworms Help the Soil?

1. **Chomping Away:** Earthworms eat fallen leaves, tiny roots, and other decaying matter, grinding it into smaller pieces.

2. **Rich Castings:** Their waste, called castings, is packed with minerals and nutrients—an all-natural fertilizer for plants.

3. **Tunneling Trails:** As they crawl through the soil, worms create air channels that let water and oxygen in. This keeps soil loose and ready for healthy roots.

Why Is This Important?

- **Better Crops:** Gardens and farms with plenty of worms often have healthier plants, thanks to aerated soil and a steady flow of nutrients.

- **Less Waste:** Worms recycle organic debris that might otherwise pile up, turning it into something valuable instead of letting it rot on the surface.

- **Eco-Balance:** Earthworms play a key role in ecosystems

around the world, supporting insects, birds, and other animals that rely on worm-rich soil.

Fun Facts

- Earthworms can have up to **five hearts**, though each heart is simpler than a human heart.
- A single earthworm can process **10 to 15 tons** of soil in one acre every year—that's a lot of recycling!
- Some species, like the **giant Gippsland earthworm** in Australia, can stretch over three feet in length.

Try It Yourself

1. **Worm Bin Compost:** With an adult's help, set up a small worm compost bin in a container. Add food scraps like fruit peels and watch worms turn them into dark, crumbly soil.
2. **Soil Test:** Dig a small hole in your yard or a nearby garden. Can you spot any worms or worm tunnels? Notice if the soil is loose and crumbly.

3. **Draw a Worm World:** Sketch an underground scene with worms creating tunnels, plants soaking up nutrients, and insects or birds feeding.

Why It Matters

Earthworms may seem unremarkable, but these wriggly recyclers help keep our soil healthy, making them **hidden heroes** of the underground world. By transforming dead leaves into rich compost and improving soil structure, they fuel plant growth and sustain countless other creatures. It's one more example that even the smallest, slimiest critters can have **super-sized impacts** on our environment!

45

Skipping Stones and Water Waves

—How Ripples Travel Farther Than You Think

Have you ever tossed a flat rock across a pond and watched it skip? Each bounce on the water's surface creates rings of **ripples** that spread out in circles. Though these waves look small, they carry **energy** outward from where the stone first touched. Believe it or not, this simple action reveals the power of waves in **lakes, oceans, and even our atmosphere**, where tiny disturbances can travel long distances!

How Do Stone Skips Create Waves?

1. **Impact and Bounce:** When the stone hits the water at a shallow angle, it forces the surface downward, then

bounces off, creating waves.

2. **Expanding Rings:** After each skip, circles of water spread outward, passing energy along to new water molecules.

3. **Energy in Motion:** The stone eventually sinks, but the waves it made keep traveling until they lose energy or reach shore.

Why Is This Important?

- **Wave Basics:** Skipping stones helps us understand how waves behave—whether it's ripples on a pond or big ocean swells.

- **Energy Transfer:** Waves move energy (not water) across distances. This concept applies to ocean waves, sound waves, and even light waves!

- **Cause and Effect:** A small action—a single splash—can lead to a big series of ripples, showing how one little disturbance can have a wide impact.

Fun Facts

- **Record Skips:** Some champion stone-skippers have achieved over 80 skips on calm waters—imagine the rings that creates!
- **Olympic Sport?** There have been playful suggestions to turn stone-skipping into a competition, with skillful throwers perfecting their aim and technique.
- **Meteorology Link:** Tiny changes in air pressure can send ripples through our atmosphere, sometimes affecting weather patterns far away.

Try It Yourself

1. **Skimming Practice:** Find a smooth, flat stone and a calm body of water (with an adult). Experiment with the angle and speed of your throw to see how many skips you can achieve.
2. **Count the Ripples:** After each skip, count how many circles form and watch how they move. Notice that they grow wider but less noticeable as they spread.

3. **Draw a Ripple Map:** Sketch a pond with a stone hitting the water. Draw circles getting bigger and label them as traveling waves of energy.

Why It Matters

Skipping stones may seem like child's play, but it teaches us about **waves**—a force shaping Earth's seas, weather, and even outer space! From a pebble's gentle drop to massive ocean swells, every wave starts with a single action. It's a lesson that **small causes can have broad effects**, reminding us that sometimes the littlest pebble can set great ripples in motion.

46

Lichens That Clean the Air

—Fuzzy Growths on Rocks Telling Us About Pollution

Have you ever noticed patches of fuzzy, crusty, or leafy growths on rocks, trees, or walls? These are **lichens**, a unique partnership between a fungus and an algae (or sometimes a fungus and a bacterium). While they might look like simple splotches of color, lichens actually do big things—like **cleaning the air** and giving scientists clues about air pollution levels. They're nature's living filters!

How Do Lichens Work?

1. **Fungus + Algae:** In a lichen, the fungus provides structure and protection, while the algae (or bacteria)

produces food through photosynthesis.

2. **Air Quality Monitors:** Lichens absorb water, nutrients, and pollutants straight from the air. If pollution is high, lichens may stop growing, change color, or die off.

3. **Tiny Filters:** As they soak up pollutants, lichens help clear the air of harmful substances—though too much pollution can harm them, too.

Why Is This Important?

- **Pollution Warnings:** By observing where lichens grow (and where they don't), scientists can tell if the air is safe or polluted.

- **Habitat Builders:** Lichens can live in tough places, like bare rocks, paving the way for other organisms by slowly breaking down the surface into soil.

- **Oxygen Creators:** The photosynthetic algae in lichens release oxygen, just like plants do, contributing a bit to our breathable air.

Fun Facts

- Lichens can survive in extreme conditions, from Arctic tundra to hot deserts—and even on old brick walls in cities!
- Some lichens glow under UV light, revealing hidden patterns and shapes.
- A single lichen is actually multiple species living together—a perfect example of nature's teamwork.

Try It Yourself

1. **Lichen Hunt:** Look for lichens on tree trunks, rocks, or fences around your home or a local park. Notice the variety of colors—green, orange, yellow, grey.
2. **Photo Journal:** Take pictures (with an adult) of different types of lichens and see if they are more common in areas with clean air (like rural places) than busy city streets.
3. **Lichen Drawing:** Sketch a lichen close-up. Label the fungus part and the algae part, showing how they fit

together.

Why It Matters

Lichens might seem like odd patches growing in forgotten corners, but they're actually critical bioindicators of our planet's well-being. By quietly filtering pollutants and transforming rough surfaces into livable habitats, lichens link air quality, biodiversity, and environmental health. They're another reminder that even the smallest, most overlooked forms of life can play **huge** roles in keeping our world clean and thriving.

47

Deep-Sea Vents and New Life

—Hot Volcanoes Underwater Feeding Strange Creatures

Deep below the surface of the ocean, where sunlight can't reach, **volcanic vents** spew super-hot, mineral-rich water into the icy darkness. These **hydrothermal vents** look like towering chimneys releasing cloudy, scalding fluids. Surprisingly, they're home to bizarre worms, giant clams, and bacteria that thrive in what seems like an impossible place for life. It's a hidden world of heat and mystery—one that reveals how adaptable Earth's creatures can be.

How Do Deep-Sea Vents Support Life?

1. **Hot, Mineral-Rich Water:** Water heated by magma

underground bursts out of cracks in the seafloor, carrying minerals such as sulfur and iron.

2. **Chemoautotrophic Bacteria:** Instead of using sunlight for energy (like plants on land), special bacteria use chemicals from the vents to make food.

3. **Symbiotic Relationships:** These bacteria often live inside tube worms or clams, providing nutrients to their hosts in exchange for a safe home.

Why Is This Important?

- **Alternate Life Systems:** Deep-sea vents prove that life can exist without sunlight, relying on chemical energy—challenging what we thought life needed to survive.

- **Evolution Clues:** Studying vent creatures may help scientists understand how life on Earth began billions of years ago—or how it might exist on other planets or moons.

- **Ocean Balance:** These unique ecosystems recycle minerals and support food webs in the darkest parts of the sea, connecting distant parts of Earth's oceans.

Fun Facts

- The water escaping these vents can reach **700°F (370°C)**—hot enough to melt some metals!
- Tube worms near vents can grow to over **6 feet** long in just a few years, thanks to their bacterial partners.
- Some scientists think similar vents could exist beneath the ice of Jupiter's moon Europa or Saturn's moon Enceladus, possibly harboring alien life.

Try It Yourself

1. **Vent Exploration:** Look up real photos or videos of hydrothermal vents (with an adult). Notice the chimney-like structures and unusual animals, like tube worms or ghost-white crabs.
2. **Alien World Drawing:** Sketch a deep-sea vent scene. Add black "smoke," weird creatures, and label which ones rely on chemosynthetic bacteria for food.

3. **Imagine Other Planets:** Discuss with family or friends how hydrothermal vents on other worlds might look. Could alien worms or bacteria survive there?

Why It Matters

Hydrothermal vents remind us that our planet has **many** hidden realms where life has adapted in astonishing ways. Even in the crushing pressure and scorching heat of the deep ocean, communities of creatures flourish. These vents expand our ideas about what's possible and hint that life could appear in the unlikeliest places—even beyond Earth's blue seas.

48

Elephants and Underground Water

—Giant Footprints That Create Muddy Oases

When elephants stroll across the savanna, their massive feet can leave deep impressions in the soil. After a rainstorm, these footprints fill with water, creating **tiny watering holes** for frogs, insects, and birds. Even larger holes, dug by elephants searching for water, can become **mini oases** where many animals stop to drink. It's another surprising way that elephants, by simply walking and digging, change their habitat for the better!

How Do Elephant Footprints Help Others?

1. **Water Collection:** Rainwater gathers in the prints,

forming small puddles where tadpoles can grow, insects can breed, and birds can bathe.

2. **Digging Wells:** In dry seasons, elephants sometimes dig with their trunks to reach hidden groundwater. These holes later provide a drinking spot for thirsty wildlife.

3. **Ecosystem Engineering:** By reshaping the landscape, elephants create more varied environments—open paths, watering holes, and shade under fallen trees.

Why Is This Important?

- **Boosting Biodiversity:** More sources of water mean more kinds of animals can survive, from tiny bugs to larger mammals.

- **Helping Plants Thrive:** Extra water patches can encourage new plant growth, creating mini "green zones" in otherwise dry areas.

- **Keystone Species:** Like in many other ways, elephants' actions have a **big influence** on the ecosystem, making them essential for a healthy savanna or forest.

Fun Facts

- Elephants can drink up to **50 gallons** of water a day—no wonder they're experts at finding underground water!
- One elephant footprint can hold enough water for small creatures to sip on for several days.
- Elephants have sensitive trunks that can smell water below the surface, guiding them to the best digging spots.

Try It Yourself

1. **Habitat Illustration:** Draw a scene of an elephant wandering through a dry savanna. Show how footprints fill with rain, helping other creatures.
2. **Elephant Trivia:** Look up (with an adult) more ways elephants shape their environment, from uprooting trees to carrying seeds.
3. **Nature's Well:** If you ever walk on a muddy trail, look

back at your footprints after rain. Imagine how little critters might use them as tiny watering spots!

Why It Matters

Elephants are often called **ecosystem engineers** because of how drastically they reshape their surroundings—just by being elephants! Their footprints and water-digging skills provide lifelines to countless smaller species. It's yet another reminder of how one large, roaming creature can **lift up an entire community** in the wild.

49

Wind-Carved Canyons and Ancient Climates

—*Rocks Holding Weather Stories from the Past*

Have you ever walked through a **canyon** and noticed the layers of rock towering above? Over millions of years, **wind**, **water**, and **time** have shaped these cliffs into spectacular formations. But these rocky walls aren't just pretty scenery—they're **history books**, holding clues about ancient climates and weather patterns that date back long before humans ever arrived!

How Do Canyons Reveal Ancient Climates?

1. **Layer by Layer:** Each layer of sediment (sand, mud, volcanic ash) holds fossils, minerals, and other signs of what Earth's climate was like when that layer formed.

2. **Weather Markers:** Ripple marks might show where an ancient river once flowed, while certain minerals indicate a long-ago desert environment.

3. **Geologic Records:** Scientists read these layers like pages in a book, piecing together how temperatures, rainfall, and even plant life changed over time.

Why Is This Important?

- **Climate Clues:** By understanding past climate shifts, we can better predict how our planet might change in the future.

- **Fossil Discoveries:** Canyons often preserve remains of ancient animals and plants, giving us a window into evolving life on Earth.

- **Erosion Insights:** Studying how wind and water carve

canyons helps us understand erosion processes that still shape landscapes around the globe today.

Fun Facts

- The **Grand Canyon** in the United States has rock layers dating back nearly two billion years!
- Some canyon walls have visible fossils of ancient seashells, proving that land was once under the ocean.
- Wind-carved rock formations, called **hoodoos**, stand like tall pillars, formed by softer rocks eroding away faster than harder layers.

Try It Yourself

1. **Layer Cake Experiment:** With an adult, create "sediment layers" in a clear container using sand, dirt, pebbles, and different colors of soil. Each layer represents a different "era."
2. **Fossil Hunt:** Visit a local museum or fossil site to see rocks with shell imprints or plant fossils. Notice how they're found in specific layers.

3. **Canyon Sketch:** Draw a canyon wall with multiple stripes of color. Label each layer with an imaginary timeline, thinking about what creatures lived in each era.

Why It Matters

Canyons formed by wind and water are like **time machines** carved into the Earth, preserving records of changing environments, ancient creatures, and shifting climates. By "reading" these rock layers, we learn lessons that help us care for our planet today. From towering canyon rims to hidden fossils, each layer whispers the story of Earth's deep and ever-changing past.

50

Human Hearts and Whale Songs

—*Surprising Rhythms Linking Us Across the Seas*

Picture a vast ocean and a **whale** singing its haunting melody below the waves. Now think of your own **heartbeat**, thumping away in your chest. At first glance, they might seem completely unrelated. But scientists have discovered that some whale songs carry a calming rhythm that can **slow down our heart rates**, connecting these mighty ocean creatures to us in a deeply personal way.

How Do Whale Songs Affect Us?

1. **Deep, Resonant Tones:** Whale calls—especially from large whales like humpbacks—produce low-frequen-

cy sounds that can travel for miles underwater. When played above water, these tones can feel soothing to humans.

2. **Relaxing Response:** Certain studies suggest hearing these deep calls may help lower stress or heart rates, much like soft music or nature sounds do.

3. **Shared Calm:** It's as if whales and humans share a built-in response to gentle, rhythmic sounds, hinting at our basic biological connection to nature's slow, steady pulses.

Why Is This Important?

- **Stress Relief:** Listening to natural sounds like whale songs can help people unwind, supporting mental health and well-being.

- **Nature Bond:** Feeling our hearts respond to a whale's call reminds us that we're part of Earth's living network, intertwined in ways we're still discovering.

Fun Facts

- Humpback whales can sing "songs" that last over 20 minutes, repeating complex patterns that evolve each season.
- Blue whales have calls so loud they can be heard hundreds of miles away—yet the deep pitch can also be strangely comforting.
- Some whale songs change "styles" each year, almost like a new hit single sweeping across ocean waters!

Try It Yourself

1. **Sound Experiment:** Look up (with an adult) recordings of whale songs. Listen with your eyes closed and see if you notice your breathing or heart rate slowing.
2. **Nature's Playlist:** Create a calming playlist that includes ocean sounds, bird calls, or whale songs. Use it when studying, relaxing, or before bedtime.

Why It Matters

Our planet may be huge, but tiny connections—like a comforting rhythm that soothes both whales and people—reveal a deep connection. Whale songs remind us that **we're never truly alone** on this Earth; our hearts can literally beat in time with life under the waves. It's one more reminder that the threads binding us all—humans, animals, land, and sea—are far more intertwined and wondrous than we ever imagined.

51

Our Amazing Web of Life

—How Every Discovery Connects Us All

Throughout this book, we've seen how the tiniest bugs, the largest creatures, and even invisible forces can link places and beings around the globe. From **cosmic dust** falling in your backyard to **whale songs** that soothe our hearts, the world is brimming with surprising stories of connection.

Have you noticed a pattern?

Everything—from towering volcanoes to hidden mushroom networks—plays its part in a grand puzzle. Each chapter showed us that even **small actions** can lead to **big effects**, and that no piece of nature stands alone. Your own breath, your backyard, and your curiosity tie you to rainforests, deserts, oceans, and beyond!

Ready to keep exploring?

- Stay curious: Ask "Why?" whenever you see something unusual or learn a new fact.
- Share what you learn: Tell friends and family about cool discoveries, so we can all celebrate these amazing connections together.

Thanks for joining us on this exciting journey of mind-blowing connections. Our world is a place where **every living thing** is part of one amazing web of life—a web that includes **you**, weaving everyone and everything together in countless ways we've only just begun to understand. Keep learning and exploring!

For additional educational resources, please visit LeadWithGold.com.

Make a Difference with Your Review

You've made it to the end of the book—and helped spark curiosity, wonder, and deeper thinking in a young mind (maybe even your own).

I hope this book made science fun, surprising, and truly unforgettable—and helped connect the dots between everyday moments and our amazing, interconnected world.

If it inspired learning, laughter, or those wide-eyed "wait, what?!" moments, I have one small request: would you consider passing it on to other curious families and educators by taking just one minute to leave a review?

Why Your Review Matters—It's Simple, Fast, and Makes an Impact

Most parents and teachers rely on reviews to choose educational books. Your words could help:

- One more child discover that science is cool, weird, and wildly connected
- One more parent or teacher find a book that actually

gets kids excited about learning

- One more family create screen-free moments of discovery and connection
- One more curious kid realize the world is more magical—and more connected—than they ever imagined

Your review might be the very reason someone gives this book a try—and lights a spark that could last a lifetime.

How to Leave a Review

It only takes one or two sentences—share what your favorite chapter was or how your child reacted.

To leave your review, please scan the QR code below or visit: https://www.amazon.com/review/review-your-purchases/?asin=B0F26DDR35

Thank You for Paying It Forward

I hope this book helped bring wonder and connection into your world—and I thank you for helping others discover the same joy.

With deep gratitude,

Keryn Gold, PhD, MBA

Author, *Mind-Blowing Connections for Kids!*

LeadWithGold.com

About the Author

—Defying Artificial Limits & Empowering People To Achieve The Seemingly Impossible

Keryn Gold, PhD, MBA is an award-winning, globally recognized executive strategist, consulting leader, scientist & medical educator.

Drawing on her genuine drive to empower others, extensive expertise in AI, science, & leadership, her clients consistently report she's enabled them to achieve extraordinary results phenomenally fast, without overwhelm.

Keryn has run companies and led data science, strategy, & business transformation at multiple Fortune 50, startup, & FAANG companies across industries, consistently delivering 50x+ first-year ROI—without layoffs or reductions in product or service quality—& has been featured in Authority Magazine,

CXO Outlook, International Business Times, LA Weekly, & Disrupt Magazine, among others.

Her mission? To empower current & future leaders to unlock their full potential, achieve more than they ever thought possible (without burnout!), & overcome seemingly insurmountable obstacles in the age of AI.

For additional free resources, courses, and more, scan the QR code or click the link below: LeadWithGold.com.

Keryn's Other Books

—*Empowering You To Achieve What You Once Thought Impossible*

1. **The Leadership Playbook, Revisited:**
   - How To Accelerate Success, Guide & Empower Your People, Navigate Change, & Build Good AI Habits to Take Control of Your Work Life & Future-Proof Your Career, Today
2. **From "Misunderstood, Quiet, Introverted, Empathic, Highly Sensitive Person" to "Undeniable, Impactful, Changemaker"**
   - How to Harness Your Gifts to Do More Good & Own Your Future, Without Losing Yourself
3. **Personalized Learning Made Easy: Empowering**

Students with AI

- How-To Guide to Set Your Kids Up for Success, Improve Engagement, Boost Outcomes & Unlock Your Child's Potential With Adaptive AI Coaching

4. **Teaching & Education Without Burnout**
    - How AI Tools Empower Teachers, Future-Proof Students, Enable Innovation, Unlock Personalized Learning, Save Time, Improve Outcomes, & Reignite A Love of Learning
5. **Unlocking True Relief: Why Fascial Counterstrain Works When Physical Therapy, Massage, & Chiropractic Medicine Have Failed**
    - Undo Chronic Pain & Feel Better The Safe, Science-Backed Way Celebrities Do
6. **The Root Cause of Your Symptoms, Uncovered**
    - How Intracellular Micronutrient Deficiencies Are Missed By Doctors, But Cause Many Persistent Health Challenges-A Holistic Science-Backed

Guide To Feel Better

7. **Gut Health Breakthrough:**
    - Resolve Persistent Health Challenges, Autoimmune Disease & Fatigue By Restoring Your Microbiome-A Comprehensive Science-Backed How-To Guide to Heal Inflammation & Feel Better
8. **Mind-Blowing Connections For Kids! The Book of Interesting, Weird-But-True Fun Science Facts**
    - Why Earth, Space, Plants, Animals & The Brains of Smart, Curious Children Share Surprising Links

If you're interested in any of Keryn's other books and would like a free detailed summary, please scan the QR code below or click LeadWithGold.com/freegift

Appendix

—More Brain-Expanding Learning Platforms For Kids

1. AI Tutors

- **Khan Academy AI**
    - **What It Does**: Uses AI-powered feedback to guide students through math, science, humanities, and more.
    - **Enables Kids To**: Get instant help, personalized hints, and tailored practice recommendations.
- **DreamBox**
    - **What It Does**: Adaptive math learning platform offering real-time evaluation and auto-adjusted lesson

paths.

  - **Enables Kids To**: Progress at their own pace, filling in knowledge gaps and accelerating when ready.

- **Photomath**
  - **What It Does**: Allows users to take a picture of a math problem and see step-by-step solutions.
  - **Enables Kids To**: Understand the full process of solving math problems, not just the answer.

- **iReady**
  - **What It Does**: Diagnostic-driven instruction in math and reading; combines assessments and personalized lessons.
  - **Enables Kids To**: Target specific areas for improvement, track progress, and build strong foundational skills.

2. AI Writing

- **Grammarly AI**

  - **What It Does**: Provides grammar, spelling, and style corrections in real time.
  - **Enables Kids To**: Develop better writing habits, refine essays, emails, and creative writing pieces.

- **Quillionz**
  - **What It Does**: Uses AI to generate reading comprehension questions from any text.
  - **Enables Kids To**: Practice critical thinking, quiz themselves, and improve reading comprehension.

- **Jasper AI (Writing & Public Speaking)**
  - **What It Does**: Generates text and speaking prompts; can also suggest outlines and ideas.
  - **Enables Kids To**: Brainstorm effectively, develop public-speaking scripts, and polish writing projects.

3. AI Coding

- **Replit AI**

- **What It Does**: Online coding environment with AI assistance, offering real-time code suggestions.
- **Enables Kids To**: Learn multiple programming languages collaboratively, debug quickly, and build projects with AI guidance.

- **Code.org**
  - **What It Does**: Interactive lessons teaching kids basic programming concepts with block-based and text-based activities.
  - **Enables Kids To**: Progress from beginner-friendly coding tutorials to more advanced topics, encouraging creativity.
- **Scratch AI**
  - **What It Does**: Visual drag-and-drop programming environment with AI-inspired blocks and extensions.
  - **Enables Kids To**: Create interactive stories, animations, and games while exploring foundational AI

concepts.

4. AI Creativity

- **Ideogram & DALL·E**
    - **What It Does**: AI model that generates images from text prompts.
    - **Enables Kids To**: Experiment with digital art, explore visual storytelling, and expand creative expression.
- **Runway ML**
    - **What It Does**: Provides a suite of AI-powered tools for video editing, image generation, and other creative tasks.
    - **Enables Kids To**: Produce multimedia projects, learn advanced editing techniques, and explore innovative visual effects.
- **StoryLab AI**
    - **What It Does**: AI tool for generating story plots,

character outlines, and creative writing prompts.

- **Enables Kids To**: Jump-start fiction writing, enhance narrative structure, and develop storytelling skills.

- Gamma AI
    - **What It Does**: AI tool that generates compelling presentations that bring ideas to life.
    - **Enables Kids To**: Put together amazing presentation decks to showcase their ideas.

5. AI Public Speaking

- **Yoodli AI**
    - **What It Does**: Analyzes recorded speeches or presentations to offer feedback on pacing, filler words, and clarity.
    - **Enables Kids To**: Gain confidence, refine delivery, and track progress in real-time speech performance.
- **Speeko AI**

  - **What It Does**: Uses speech recognition to assess clarity, intonation, and word choice.
  - **Enables Kids To**: Practice in a safe environment, receive guided practice exercises, and build public-speaking skills.
- **Jasper AI (Speech)**
  - **What It Does**: Creates draft scripts for speeches or presentations and provides suggestions for improvements.
  - **Enables Kids To**: Polish speaking engagements, structure logical arguments, and practice tone and pacing.

6. AI for Parents

- **Google Read Along**
  - **What It Does**: Uses speech recognition to guide children through reading activities, offering corrective feedback.

    - **Enables Kids To**: Improve reading fluency and pronunciation while parents can monitor progress.

- **Duolingo AI**
    - **What It Does**: Employs AI to tailor language-learning lessons; covers a wide array of languages.
    - **Enables Kids To**: Practice language skills in bite-sized gamified lessons and stay motivated via progress tracking.

- **Wysa AI**
    - **What It Does**: AI-driven "chat buddy" for mental health check-ins and emotional well-being.
    - **Enables Kids To**: Build resilience, share feelings, and develop coping strategies with supportive prompts.

7. Comprehensive E-Learning Platforms

- **BrainPOP & BrainPOP Jr.**
    - **What It Does**: Offers short, animated videos on

science, social studies, math, and more for K–8.

- **Enables Kids To**: Engage visually, build early content knowledge, and follow up with interactive quizzes.

- **Khan Academy (Main Site)**
  - **What It Does**: Free lessons in math, science, humanities, test prep, and more for all ages.
  - **Enables Kids To**: Access an entire library of self-paced courses and practice exercises with detailed feedback.
- **Khan Academy Kids**
  - **What It Does**: Free program for ages 2–7 focusing on early literacy, reading, writing, language, and math.
  - **Enables Kids To**: Develop foundational reading and math skills, stay engaged with interactive characters, and track progress.
- **IXL**

  - **What It Does**: Offers a range of academic practice exercises in math, language arts, science, and social studies.

  - **Enables Kids To**: Receive immediate feedback, track skill mastery, and practice targeted areas by grade level.

- **Adventure Academy**

  - **What It Does**: Immersive online environment covering math, language arts, science, and social studies (ages 8–13).

  - **Enables Kids To**: Explore virtual worlds, complete educational quests, and engage in gamified learning.

- **ABCmouse**

  - **What It Does**: Online curriculum (ages 2–8) covering language arts, math, science, social studies, and art through gamification.

  - **Enables Kids To**: Follow a structured learning path, earn rewards, and gain familiarity with core subjects at an early age.

- **Prodigy**
    - **What It Does**: Fantasy-themed math platform (ages 6 and up) with adaptive difficulty and battle-based gameplay.
    - **Enables Kids To**: Stay motivated through adventure-style play, earn rewards, and improve core math skills.
- **PBS KIDS / PBS Parents Play and Learn**
    - **What It Does**: Provides a wide variety of games, videos, and parent-child activities aligned with PBS KIDS TV programs.
    - **Enables Kids To**: Engage with familiar characters, strengthen parent-child interaction, and explore STEM, reading, and social-emotional skills.
- **Epic**
    - **What It Does**: Digital library with 40,000+ e-books for kids up to age 12.

- **Enables Kids To**: Develop a love of reading, access an extensive collection anytime, and track reading progress.

- **Starfall**
  - **What It Does**: Focuses on early literacy and math (Pre-K–3), featuring phonics games and reading exercises.
  - **Enables Kids To**: Learn foundational reading skills, practice spelling, and build confidence in basic math.

- **National Geographic Kids**
  - **What It Does**: Interactive platform (ages 4–12) covering animals, geography, and environmental science.
  - **Enables Kids To**: Explore the natural world via videos, quizzes, and articles—fueling curiosity.

- **Teach Engineering**
  - **What It Does**: Offers hands-on STEM activities for K–12, from simple projects to more advanced engi-

neering tasks.

- **Enables Kids To**: Apply math/science in real-world contexts, build problem-solving skills, and experiment with design thinking.

- **NASA Kids' Club**
  - **What It Does**: Collection of space-themed games and activities (ages 5–8) from NASA.
  - **Enables Kids To**: Discover basic astronomy, aerospace concepts, and spark curiosity about the universe.
- **Ted-Ed**
  - **What It Does**: Curated educational videos and lessons on various topics, from history to science.
  - **Enables Kids To**: Explore engaging mini-lectures, take interactive quizzes, and deepen understanding through rich animations.

8. Coding & STEM Exploration

- **Blockly (MakeWonder.com/apps/blockly)**
  - **What It Does**: Visual programming interface that pairs with Wonder Workshop robots (like Dash & Dot).
  - **Enables Kids To**: Learn coding logic, problem-solving, and robotics in a hands-on, fun environment.
- **Scratch Jr.**
  - **What It Does**: Simplified version of Scratch for younger children (ages 5–7), teaching block-based coding basics.
  - **Enables Kids To**: Create animations, learn sequencing, and practice computational thinking.

9. Music

- **Yousician**
  - **What It Does**: Interactive tool teaching piano, gui-

tar, bass, or ukulele via real-time feedback on playing.

- **Enables Kids To**: Develop musical skills in a fun, game-like environment with auto-assessment and progress tracking.

How to Use This Appendix

- **Finding the Right Tool**: Decide which subject or skill is a priority (e.g., math remediation, early reading, AI coding, or creative arts).
- **Combining Apps**: Mix and match to cover multiple skill areas—e.g., IXL for math and reading plus Blockly for coding.
- **Parental Involvement**: Tools labeled "AI for Parents" or "PBS Parents" can help families guide kids' learning journeys with data insights, mental health support, or co-play activities.
- **Monitoring Progress**: Many of these apps include dashboards or progress trackers, so parents and educators can watch how kids evolve over time.

This curated list showcases an array of modern educational and AI-powered platforms, each designed to engage children at their level, provide personalized feedback, or spark new interests to help them discover their passions.

References

1. Nasser, L. (Host). (2020). Connected: The Hidden Science of Everything [TV series]. Netflix. https://www.netflix.com/title/81031739

2. Encyclopedia of Life. https://eol.org

3. Intergovernmental Panel on Climate Change. https://www.ipcc.ch

4. International Union for Conservation of Nature. https://www.iucn.org

5. National Aeronautics and Space Administration. NASA climate. https://climate.nasa.gov

6. National Geographic Society. Environment.https://www.nationalgeographic.com/environment

7. National Oceanic and Atmospheric Administration. https://www.noaa.gov

8. National Park Service. https://www.nps.gov

9. Smithsonian Institution. Ocean portal.https://ocean.si.edu

10. U.S. Geological Survey. https://www.usgs.gov

11. World Wildlife Fund. https://www.worldwildlife.org

Made in the USA
Monee, IL
04 November 2026

67378728R00134